SCIENCE 101

CHEMISTRY

Denise Kiernan and
Joseph D'Agnese

Collins

An Imprint of HarperCollinsPublishers

CONTENTS

Funnels attached to flasks in a laboratory.

SCIENCE 101

CHEMISTRY

HarperCollins books may be purchased for educational, business, or sales promotional use. For information, please write: Special Markets Department, HarperCollins Publishers, 10 East 53rd Street, New York, NY 10022.

Produced for HarperCollins by:

Hydra Publishing
129 Main Street
Irvington, NY 10533
www.hylaspublishing.com

Library of Congress Cataloging-in-Publication Data

Kiernan, Denise.
Chemistry / Denise Kiernan and Joseph D'Agnese.
p. cm. — (Science 101)
Includes index.
ISBN 978-0-06-089138-1
1. Chemistry. I. D'Agnese, Joseph. II. Title. III. Series.

QD31.3.K54 2007
540—dc22

2007023024

08 09 10 QW 10 9 8 7 6 5 4 3 2

Chemists research and develop state-of-the-art pharmaceuticals to treat disease and increase life expectancy.

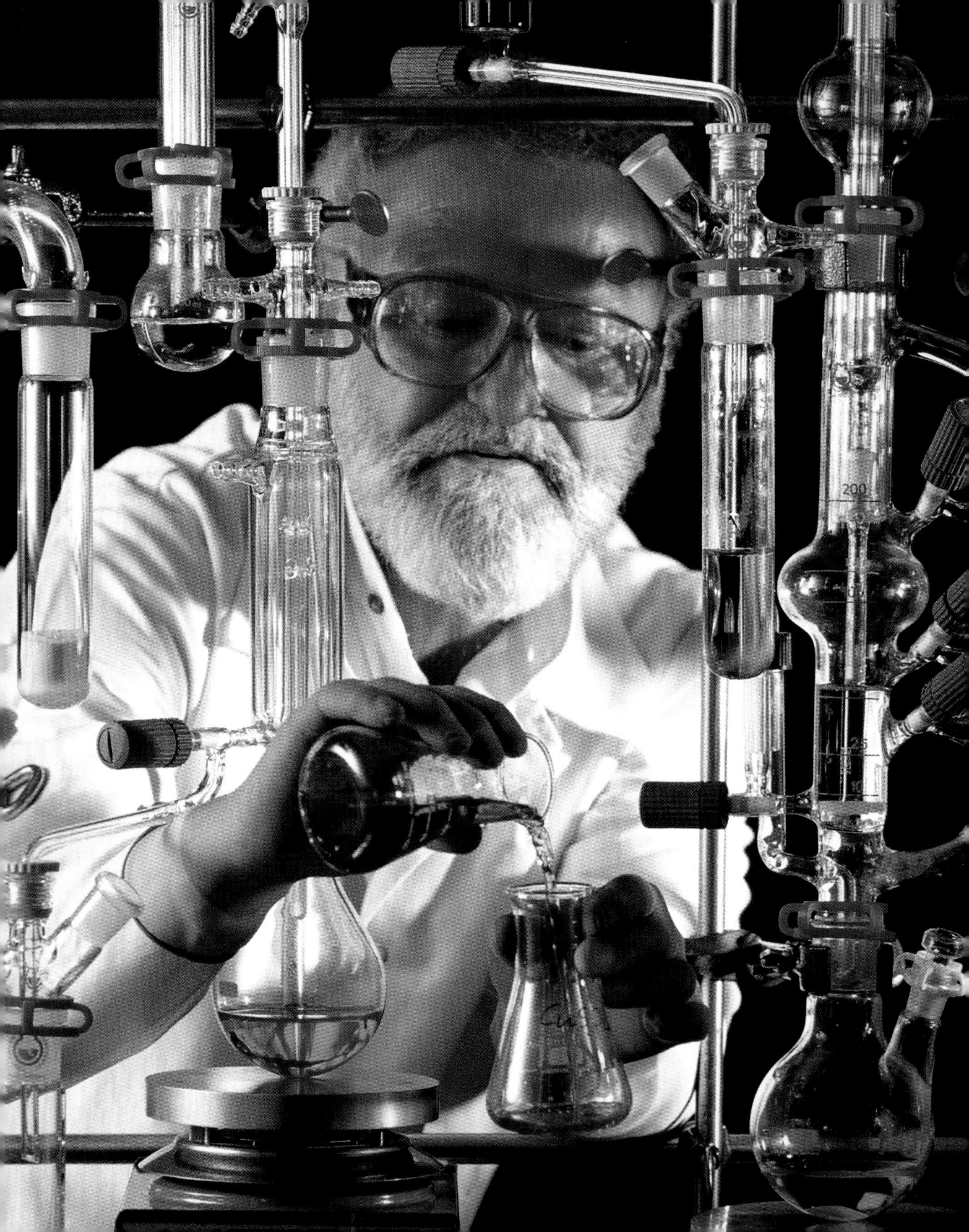
200

WELCOME TO CHEMISTRY

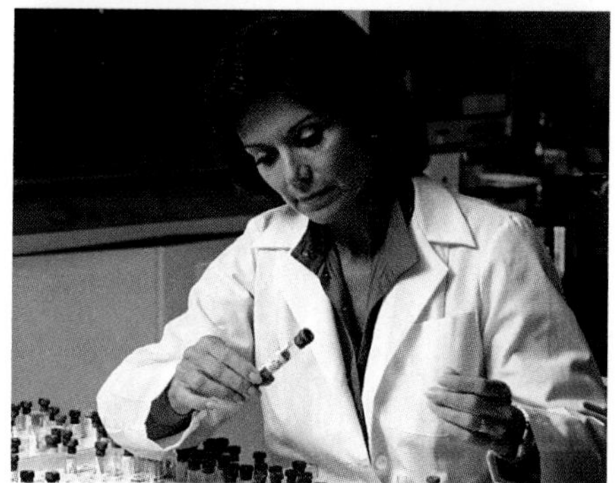

Left: The world of chemistry is complex, but scientists are constantly making new and exciting discoveries. Top: An artist's rendering of bonded atomic particles. Bottom: A scientist putting chemical samples into labeled beakers.

Curiosity. Creativity. Persistence. Humans throughout time have been driven by the need to understand and make sense of their surroundings. Our natural curiosity has led us to explore our world and those worlds beyond the boundaries of our own atmosphere, and to develop and create theories that seek to explain wonders that might otherwise be taken for granted on a daily basis. Primitive man observes that a spearhead left exposed to the elements changes in appearance over time. A mishap next to a campfire results in a shiny new substance that had not previously existed. The innate desire to express oneself spawns the creation of paints and pigments derived from plants and the earth.

This inquisitiveness fostered a methodical, scientific approach to investigations as early humans sought to delve into nature's mysteries with any means available to them. Today our tools may be very different, but our impulses are the same. Chemistry has proved to be a powerful bridge between the physical and biological sciences, uniting the organic and inorganic worlds. Physicists may be concerned with the birth of stars and planets, but it is chemistry that provides the foundation for understanding the elemental makeup of the universe. Biologists probe the workings of the human cell, but the molecules involved in these processes must be understood from within their chemical context. Chemistry is about understanding the composition and mechanisms that lie beneath the veneer of the world visible to the naked eye. And the more answers we find, the more questions we create.

ABANDONING THE PHILOSOPHER'S STONE

Ancient chemistry had its roots in philosophy. Aristotle, Democritus, and others yearned for a cogent explanation for the way the world worked, and used their gifts of logic to "reason" answers that would satisfy them and those who looked to them for answers.

The ancients marveled at expanses of white sand that appeared unified and smooth from a distance but were, in fact, composed of billions of tiny, seemingly uniform particles. This simple observation led some of science and philosophy's earliest minds to ponder the composition of matter and consider the possibility that a living being, a rock, or a palm frond might also be composed of billions of tiny units that could not be viewed from a distance. The ancient belief that all matter was formed of earth, air, fire, and water—combined with a desire to create gold from less noble metals—drove scientific thought for centuries. As science advanced, and the Age of the Enlightenment shaped the chemical revolution of the eighteenth century, the magic and mysticism that had permeated science was replaced by a commitment to empirical research and experimentation—the identification of the elements and advances in the deciphering of that building block of all known matter, the elusive atom, once called *atomos* by the Greeks so many thousands of years before.

A PRECARIOUS BALANCE

With great knowledge and power comes great responsibility. Some of the most remarkable advances in chemical science have resulted in dangerous consequences that threaten us and the world as we know it. Foolhardy hubris has the capacity to derail our best intentions, and the speed with which new discoveries can be turned into marketable products is dizzying but unsettling. In our rush to "fix" a perceived problem, we can create greater ones if the ramifications of

Biochemistry is certainly not a new field. Man has always sought ways to improve health by studying how the chemical properties of various substances affect the body.

Left: A plane spreads pesticides over a field in order to protect valuable crops from destructive insects. Below: A worker in a pharmaceutical factory mixes chemicals that will eventually be used to make medicine.

our new discoveries are not fully understood or even explored before they are put into use. The development of the pesticide DDT, for example, promised to eradicate the threat of dangerous insects. Yet even today, we are still living with its toxic legacy, one that has decimated the world's avian population and polluted our own bloodstreams. Rachel Carson, whose landmark book *Silent Spring* launched the modern environmental movement, explained it best: "For the first time in the history of the world, every human being is now subjected to contact with dangerous chemicals, from the moment of conception until death."

However, if we tread carefully, chemistry can prove to be as powerful an ally as anyone could wish for. Rational drug design holds the promise of a cancer- and AIDS-free future. Green technology is providing options that will help all of us combat some of the environmental ills that modern times have helped create. And nanotechnology will enable us to do work and provide therapies on a scale rarely even contemplated before the 1990s. As we ponder our place in an ever-expanding universe and marvel still at the billions of tiny grains of sand that line our shores, chemistry will carry our curiosity and creativity into whichever future we choose.

CHAPTER 1

ATOMS AND ELEMENTS

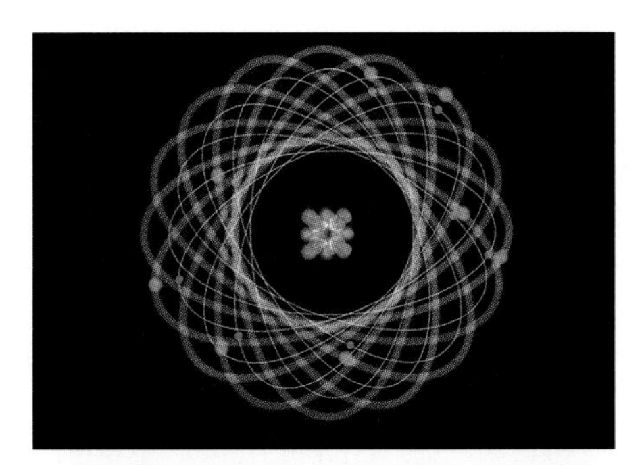

Left: The Great Salt Lake in Utah. An average grain of sodium chloride (salt) contains over two million atoms. Top: Computer artwork of atomic structure. Bottom: The atomic structure of rock changes very slowly, allowing the Earth to hold clues to life in past millennia. The remains of dinosaurs have been found in the badlands of Alberta, Canada.

The fascinating science of chemistry is made even more intriguing by the fact that it is based on something no one has ever seen—an atom. Each atom is built from three kinds of pieces—electrons, protons, and neutrons—with the number of protons strictly prescribed and the number of neutrons varying only within a limited range. The electrons are systematically situated in mathematically defined regions of space, and they make transitions between positions by acquiring precisely predictable amounts of quantum energy. All atoms in all materials are in constant motion, but while an atom in a solid may require millennia to move a millimeter, atoms in air race around at incredible speeds. Some atoms have estimated lifetimes as long as the current age of the universe, while others exist for only billionths of a second.

All the complex materials of life are built of atoms and the compounds that they form. The journey from the discovery of the atom to the development of atomic theory and beyond is one of the great stories of science.

A Theory of the Atom

Of the ancient Greek, Asian, African, and Arabic philosophers whose ideas on atoms have survived, one of the best known is Democritus, a Greek philosopher of c. 460–370 BCE. As did many of his contemporaries, Democritus believed the world could be understood by the application of logic. As such, these thinkers inspired the rational approach to science, but they eschewed experimentation, believing rigorous logic to be the purer form of investigation. Democritus thought of the atom as being the smallest particle of all matter, but he also thought that all atoms were the same. Aristotle (384–322 BCE) believed that four elements—air, earth, water, and fire—were present in all materials. The modern concept of a large variety of elements that can exist as separate, pure materials was the inspiration of Antoine Lavoisier (1743–94), though the modern concept of an atom, the smallest unit of an element, had to await the work of John Dalton (1766–1844). Dalton is given credit for our understanding of the atom because he not only offered the idea that atoms of different elements were distinguishable by characteristic mass, but grounded his theory in experimental results. With compounds such

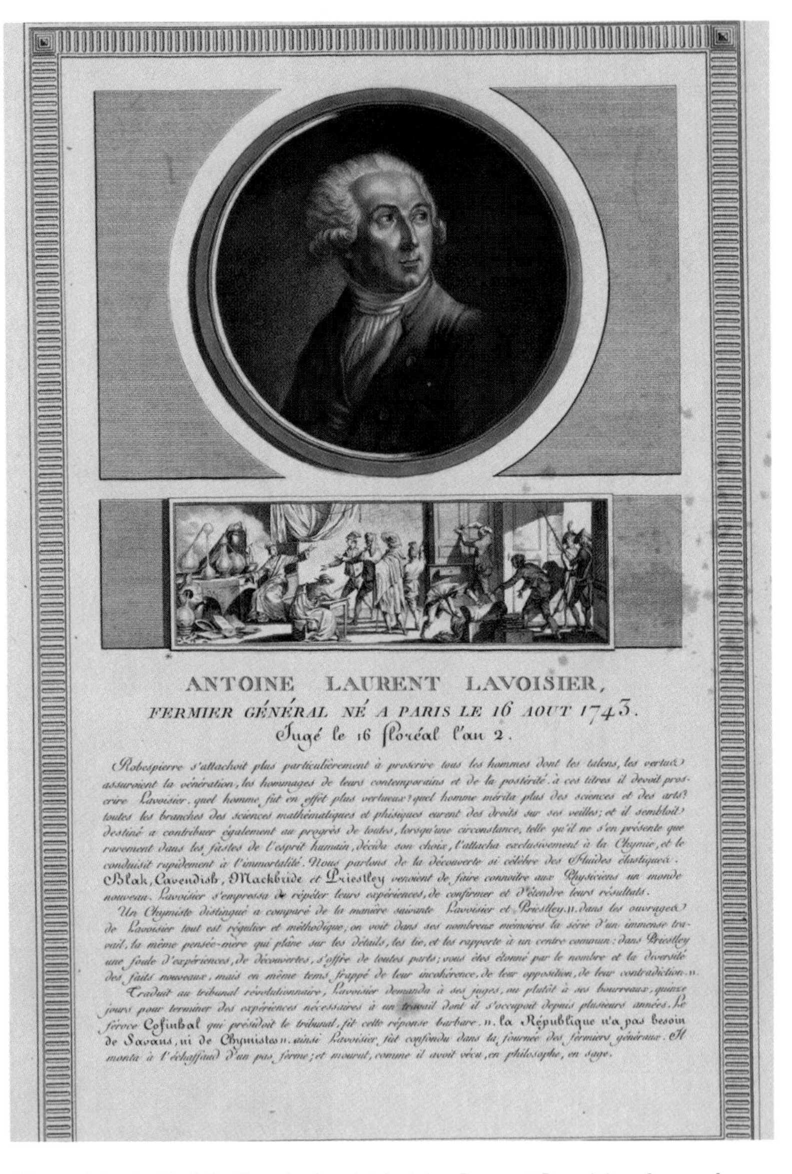

Above: A portrait of the French chemist Antoine-Laurent Lavoisier, shown above a vignette depicting his arrest and conviction in 1794. Lavoisier was beheaded during the French Revolution. Top left: All matter is made up of atoms, including that found in nebulas within our galaxy.

as carbon monoxide and carbon dioxide, Dalton was able to show that the masses of the same element in different compounds stood in whole-number ratios to each other, indicating the element came in discrete mass packets. Dalton's "ultimate particles"—atoms—explained this observed law of multiple proportions and set the stage for a closer look at the atom.

CLASSICAL ATOMIC THEORIES

The atomic theory of matter advanced by Dalton served to answer many questions concerning the behavior of chemical reactions, but these solutions always gave rise to new questions—and so it was with atoms. It became apparent by the early 1800s that the atoms themselves had structure, and by the early 1900s, positively charged nuclei and negatively charged electrons had been identified in the atom; but how these pieces fit together remained to be determined. British investigator J. J. Thomson (1856–1940), who may be credited with finding the persuasive evidence for the electron, thought that the negatively charged electrons might be embedded in a sea of positive charge, a concept that became known as the plum-pudding model because the electrons resembled the negative-charged raisins in a sea of positive-charge plum pudding, a British dessert favorite. Ernest Rutherford (1871–1937), however, in a ground-breaking experiment, fired small, positively charged particles at the atoms in a thin gold foil and saw some particles rebound straight back toward the source. Astonished by the result, Rutherford said it was as though he had "fired a cannonball at a tissue and found it bouncing back," but he correctly interpreted his results to mean that the atom has a very tiny, dense, positively charged nucleus, and the electrons are located outside the nucleus. The areas of space occupied by the electrons have come to be called orbitals.

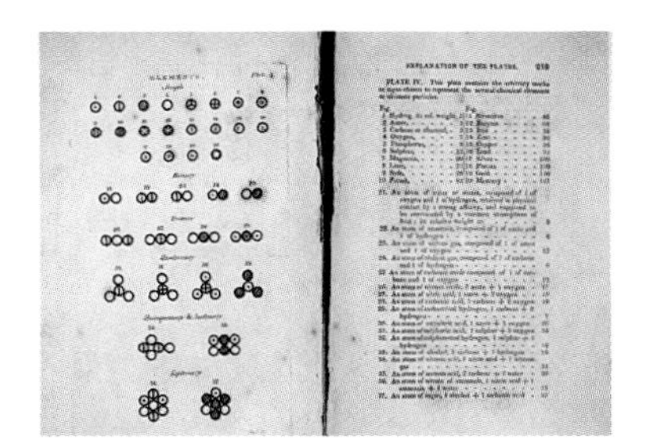

Above: Pages from John Dalton's three-volume work titled A New System of Chemical Philosophy. *The description at right explains that the symbols represent the "several chemical elements or ultimate particles." Top: Portrait of John Dalton. Left: Joseph John "J.J." Thompson.*

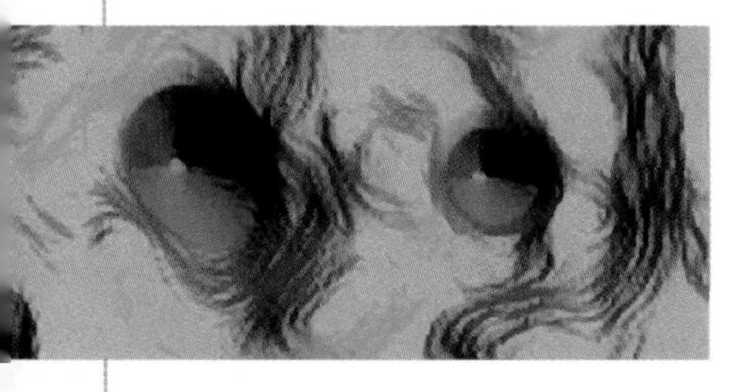

The Structure of the Atom

Two students of Rutherford, Henry Moseley (1887–1915) and James Chadwick (1891–1974), went on to show that nuclei of atoms consist of two types of particles—protons and neutrons. Individual protons are assigned one unit of positive charge, electrons have one unit of negative charge, and neutrons have no charge. Chadwick provided evidence for the uncharged neutrons by listening to the advice of his mentor, Rutherford, who said, "Look for the invisible man in Piccadilly Square by those that he pushes aside."

SUBATOMIC PARTICLES

Protons and neutrons both have a mass of 1 atomic mass unit, approximately a trillionth of a trillionth of a gram. The mass of the electron is even less. The mass of a minuscule electron compares to the mass of a proton as the mass of a tick might compare to the mass of the dog (but the tick, as we will see, determines the behavior of the dog). The identity of an element is based on the specific number of protons in the nuclei of its atoms; for instance, an atom of hydrogen must always have one, and only one, proton,

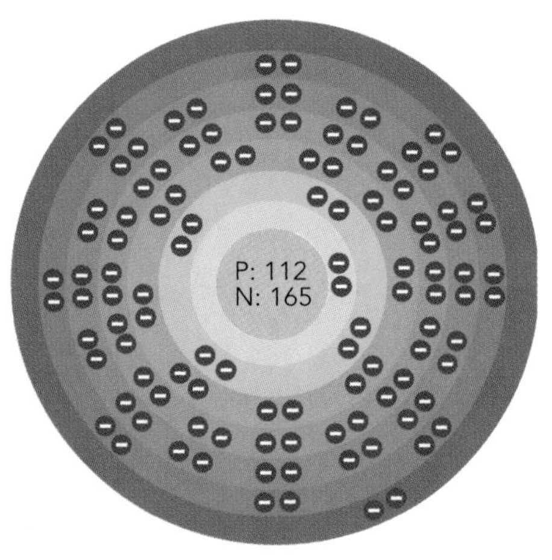

Above: The atomic number assigned to the element hydrogen (left) is 1. This number represents the number of protons a hydrogen atom possesses. The atomic number for the element ununbium (right), a transition metal, is 112. Top left: Microscopic view of cobalt atoms on a copper surface. The large round features indicate cobalt bonding to the copper at the lowest energy bonding site.

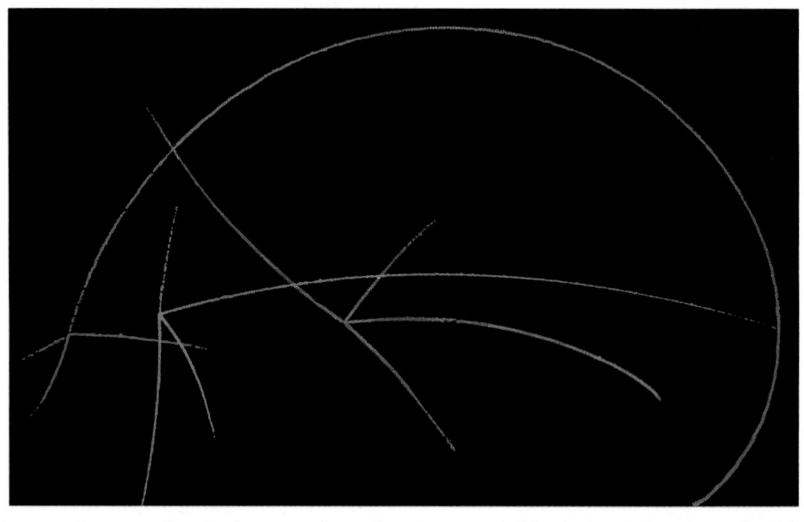

Neutrons have no electric charge and are therefore not visible in cloud chamber or bubble chamber photographs, but their presence can be betrayed by their effect on other, charged particles. In this false-color cloud chamber photo, a narrow beam of neutrons enters from below. The chamber is filled with hydrogen gas saturated with a mixture of ethyl alcohol and water.

and an atom of helium must have exactly two protons. In the modern periodic table, the elements are listed in order of increasing atomic number, which is the number of protons in an atom of that element. Reading from left to right across the periodic table, the atomic number, the whole number given with the element, begins with hydrogen, with an atomic number of 1. Helium (He) has an atomic number of 2, lithium (Li) has an atomic number of 3, and beryllium (Be) has an atomic number of 4. The numbering continues down to the most massive element thus far confirmed, with an atomic number of 114.

Scientists researching the atom in the early twentieth century found that the number of neutrons in an atom can vary, but the number of negatively charged electrons in an electrically neutral atom must equal the number of protons in the nucleus. This knowledge of protons, neutrons, and electrons completed the basic picture of the atom, but answering questions always opens the door to others, and the question now on the table was: Why doesn't the electrical attraction between the electron and the proton pull the electrons into the nucleus? Breakthroughs in modern quantum mechanics, namely the discovery that light could behave as a particle or a wave, would help provide answers.

Right: In 1908, Rutherford collaborated on the Rutherford-Geiger detector, a device that picks up changes in particle ionization and became the forerunner of the modern-day smoke detector.

Some of the physicists and chemists who contributed to the understanding of atomic structure, photographed in 1932. Seated, from left to right, are Sir James Chadwick, Hans Geiger, and Sir Ernest Rutherford. Standing behind them are, from left to right, Gyorgy Hevesy, Mrs. Geiger, Lise Meitner, and Otto Hahn. The latter two obtained the first nuclear fission seven years later.

ERNEST RUTHERFORD

The First Baron Rutherford of Nelson was born Ernest Rutherford in 1871 on a pioneer farm in New Zealand. He went from this rugged, hardscrabble life to England's Cambridge University and then McGill University in Montreal, Canada. In Canada, he carried out his famous gold-foil experiment, in which he fired small, positively charged particles at atoms in a thin gold foil and saw the rebound that allowed him to accurately describe the nucleus for the first time. When the experiment worked, he is said to have danced the *haka*, a traditional war dance of the indigenous New Zealand Maori. He won the Nobel Prize for his work in 1908.

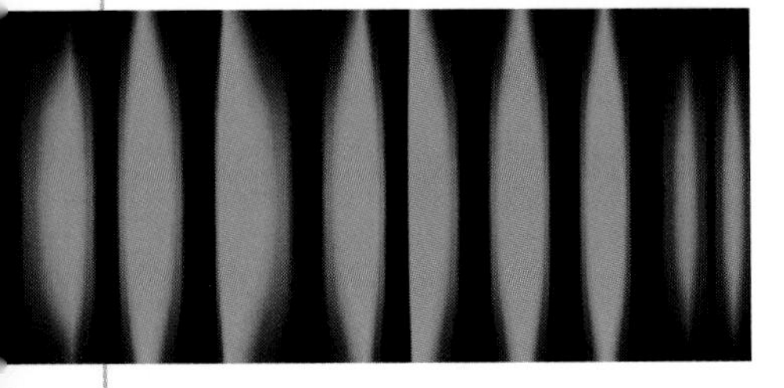

The Modern Atomic Theory

Since the first smelting ovens fired up, humans have known that the quality of light from a heated source changes with temperature. In nineteenth-century Europe, scientists understood that heat causes molecules to move, and as they move, the electric field between protons and electrons changes, oscillates, and this oscillating field is light. It also seemed reasonable that the hotter the substance, the faster molecules should oscillate and more high-frequency light should be emitted. Visible light ranges from red to violet, so scientists expected ultra-heated sources to emit light beyond violet—ultraviolet. The "ultraviolet catastrophe" stemmed from the observation that the relative amount of high-frequency light did indeed increase as the temperature increased—until a point—then it decreased.

At the very beginning of the twentieth century, German professor Max Planck (1858–1947), working on the problem of the ultraviolet catastrophe, proposed that light-generating oscillations are quantized; that is, the oscillations did not start gradually, but had to wait until they had a sufficient amount of energy, a quantum of energy, and then started all at once. Furthermore, Planck proposed, the amount of energy required to excite a particular oscillation went up as the vigor of the oscillation increased.

QUANTIZATION

Paper money is quantized: it comes in basic units of dollars. The economy is analogous to temperature: the better the economy (the higher the temperature), the more dollars are available for purchases (the more energy is available

Right: Max Planck. Top left: A computer graphic of a wave form. Light was believed to act exclusively as a wave until Max Planck and others questioned the wave theory of light in the nineteenth century.

Niels Bohr won the Nobel Prize in physics for his research on atomic structure.

to excite vibrations), and the number of purchases (oscillating molecules) goes up. But the increase in purchases is greater for the low-cost items; the sales of high-ticket items does not increase as rapidly because more money has to be accumulated before a high-price item can be purchased. So it is for Planck's oscillators. As the temperature goes up, the number of oscillators increases, but the high-frequency oscillators cost more energy to get started, so high-frequency light increases more slowly. This imbalance leads to a drop-off at high frequencies, as was experimentally observed.

MODERN QUANTUM MECHANICS

Planck thought of his quantized energy as more of a computational convenience than a physical reality. Having oscillators require specific packets of energy, quanta, to start oscillating said, in essence, that light delivered its energy all at once—acting as a particle, not a wave—and nobody wanted to throw out the wave model of light.

Quanta, however, kept showing up. Another theoretician, Niels Bohr (1885–1962), a Dane, found quanta when he tackled the question of fireworks: fireworks have different colors because atoms of elements absorb or emit only certain wavelengths of light. Bohr proposed that electrons could exist only in specific orbits, and transitions between the orbits could happen only if the precise amount of light energy necessary was absorbed or emitted: a quantum. This model proved useful for rationalizing how atoms emit light, but still did not explain why an orbiting negatively charged electron would not spiral into the positive nucleus. French physicist Louis de Broglie (1892–1987) said if light could be a particle as well as a wave, then maybe an electron could be a wave as well as a particle.

This notion of modeling a physical phenomenon as a particle under one set of circumstances and as a wave under others is not as unnatural as it seems. Sound can be modeled as a particle or a wave. When a certain pitch of sound shatters glass, it is as though the glass were hit by a bullet. Bullets do not bend around corners, though, so when sound bends around doorways and carries conversations to another room, sound is acting as a wave. After a bit of fuss, the theory that light could act as a particle or a wave and electrons could act as waves or particles became accepted. Electron waves explained why electrons do not spiral into the nucleus: waves can blend and build and self-sustain. Electron orbits were now termed orbitals to reflect their wave nature, and with this last conceptual hurdle cleared, the structure of the atom was understood. Providing confirmation, this structure was reflected in the periodic table.

Louis de Broglie proposed the idea of the dual nature of light, showing that it acts as both a particle and a wave.

arsenic 33 As 74.922	selenium 34 Se 78.96	bromine 35 Br 79.904	krypton 36 Kr 83.80
antimony 51 Sb 121.76	tellurium 52 Te 127.60	iodine 53 I 128.90	xenon 54 Xe 131.29

The Periodic Table

J. J. Thomson's electrons fit around the nucleus in three-dimensional groups of orbitals called shells. The first shell holds two electrons, the second shell holds eight electrons, the third shell holds eighteen electrons, and the fourth shell holds thirty-two electrons, and so on, in a neat mathematical progression of the shell number squared times two. The shells also include subshells, which again exist in regular patterns. The first shell has one subshell, the second shell has two subshells, the third shell has three subshells, and on it goes.

Above: Dmitri Mendeleev.
Top left: Detail showing elements of the periodic table.

As electrons are added to the elements, they fill subshells in an order dictated by the energy of the subshell, starting with the lowest energy first. On the periodic table, elements with their outermost electrons in specific subshells are grouped together. Therefore, as predicted by quantum mechanics, the number of an element's electrons, the types of subshells that hold the electrons, and certain periodic properties can be read from the periodic table.

PERIODIC PROPERTIES

Perhaps it would be more accurate to say that quantum mechanics confirms rather than predicts the shape of the periodic table, because the basic shape of the periodic table was known before the theory of quantum mechanics was in place. In 1869, Dmitri Mendeleev (1834–1907), a Russian professor of chemistry, proposed grouping the elements by chemical and physical properties and arrived at the arrangement of elements we now know as the periodic table. Though others were working on similar arrangements at the time, Mendeleev had the inspiration to leave blanks in his table where he believed undiscovered elements belonged. The subsequent discovery of these elements established the validity of Mendeleev's table.

Just as an orchestra is structured around sections of instruments, the periodic table is structured around groupings of elements. In an orchestra, the location of an individual is determined by his or her group—woodwinds, brass, strings, percussion—and whether the person is first chair, second chair, and so on. In the periodic table, an element's position is also predicated by defining properties, such as relative atomic size.

As one might expect, atomic size depends on electrons, but size is not directly related to the number of electrons. The size of the atoms of an element increases down a column of the periodic table, but actually decreases from left to right. The shrinkage from left to right is attributable to the addition of protons to the nucleus. The greater positive charge pulls in the electrons. However, the element at the end of each row represents a filled shell, and this filled shell shields the next electrons from the full force of the positive nucleus, much as an iron filing might be shielded from the full force of a magnet by a layer of lead. As a result, the atomic size increases as the atomic number increases down a column.

Element Periodic Table

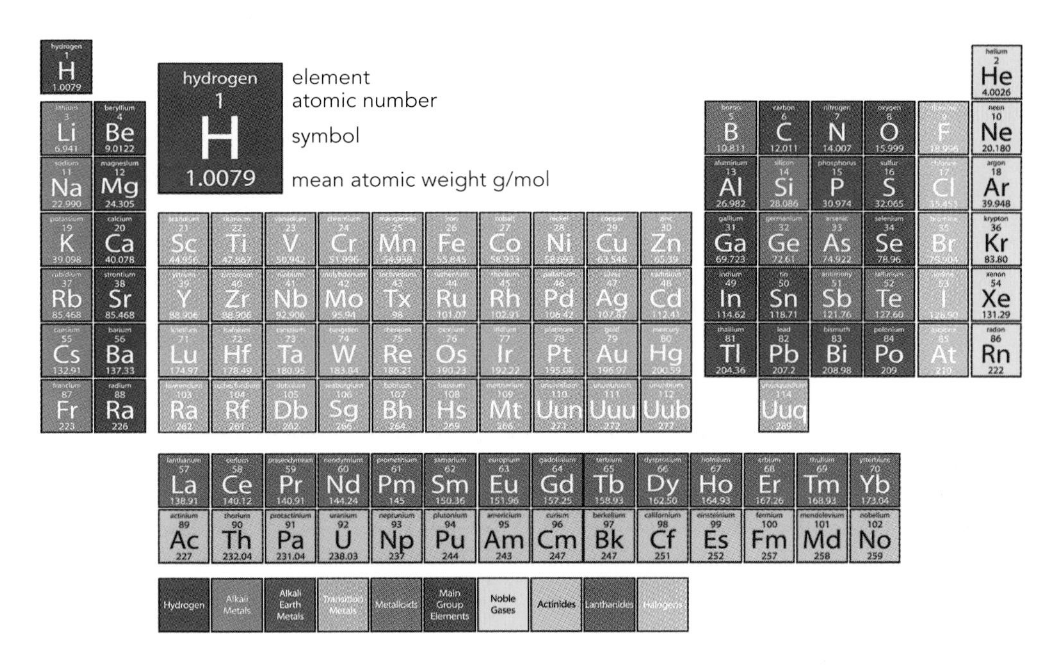

The number of electrons in an atom matches the number of protons when the atom is neutral for a reason. Atoms do not have to be electrically neutral. They can be negatively charged or positively charged ions if they gain or lose electrons. The tendency for electrons to be gained or lost, and the most likely number to be gained or lost, is another property that can be predicted by knowledge of an element's position on the periodic table. The interpretation of this periodic property, however, depends on whether an element is classified as a metal, a nonmetal, or that curious maverick, a metalloid.

An orchestra is organized into sections based on specific criteria of individuals, much as the periodic table (shown above) is structured according to the properties of each element.

Metals, Nonmetals, and Metalloids

The periodic considerations—the number of electrons around a nucleus, their distance from the nucleus, and the shell in which they reside—converge to establish a very important grouping of elements on the periodic table: metals, nonmetals, and metalloids.

The line of demarcation between metals and nonmetals on the periodic table is the dark zigzag, staircase-shaped line on the right that begins between boron (with the symbol B) and aluminum (Al).

A glance at the placement of this line reveals a decided asymmetry. By far, the majority of the elements on the periodic table are metals. Elements to the left on the periodic table, such as copper, iron, silver, and gold, tend to behave more like metals, whereas elements more to the right, such as oxygen, nitrogen, and helium, often have decidedly nonmetallic behaviors. The characteristics of pure metals are familiar from common experience; most metals (with the exception of mercury) are solids at room temperature, shiny, and good conductors of heat and electricity. A metal has these properties because, in a metal, the nuclei can be thought of as being surrounded by a sea of electrons. This closeness of the nuclei and mobility of electrons leads to their tendency to be solid at room temperature and their good thermal and electrical conductivity. The sea of electrons can also reflect visible light, lending luster to a metal surface. When metals enter into chemical reactions, they tend to lose electrons and form positive ions. The positive charge results when the protons in the nucleus outnumber electrons that remain after a chemical reaction.

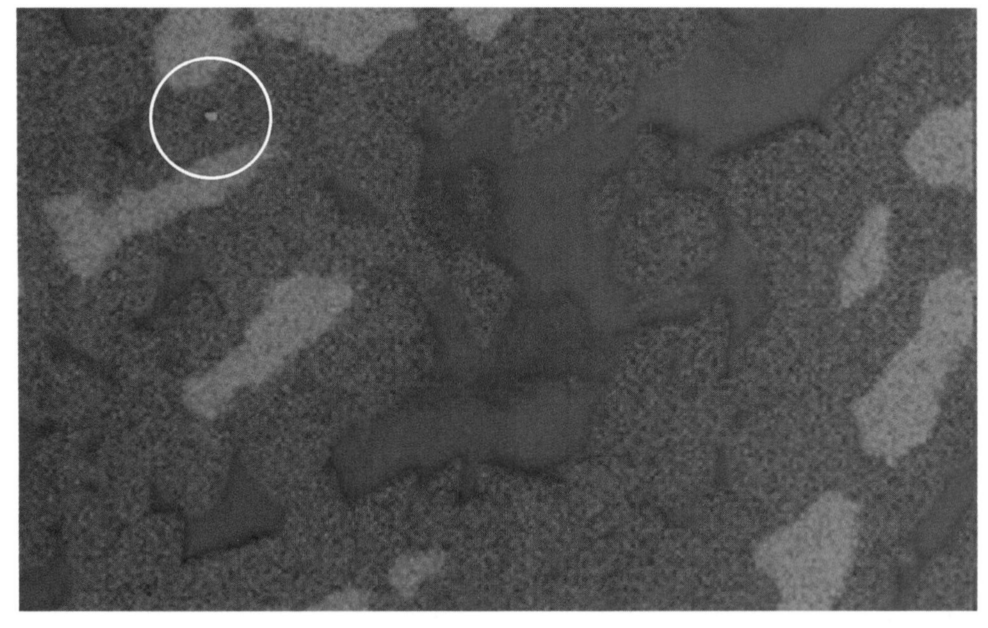

Above: This X-ray image from a scanning electron microscope shows a sample of a nickel-aluminum alloy. Pure aluminum is shown in blue, pure nickel in red, and the alloy of the two by other colors. The green dot at upper left shows a contaminant particle of chromium. Top left: Lithium reacting with water. Light enough to float, this highly reactive element ignites and burns when exposed to water in oxygen.

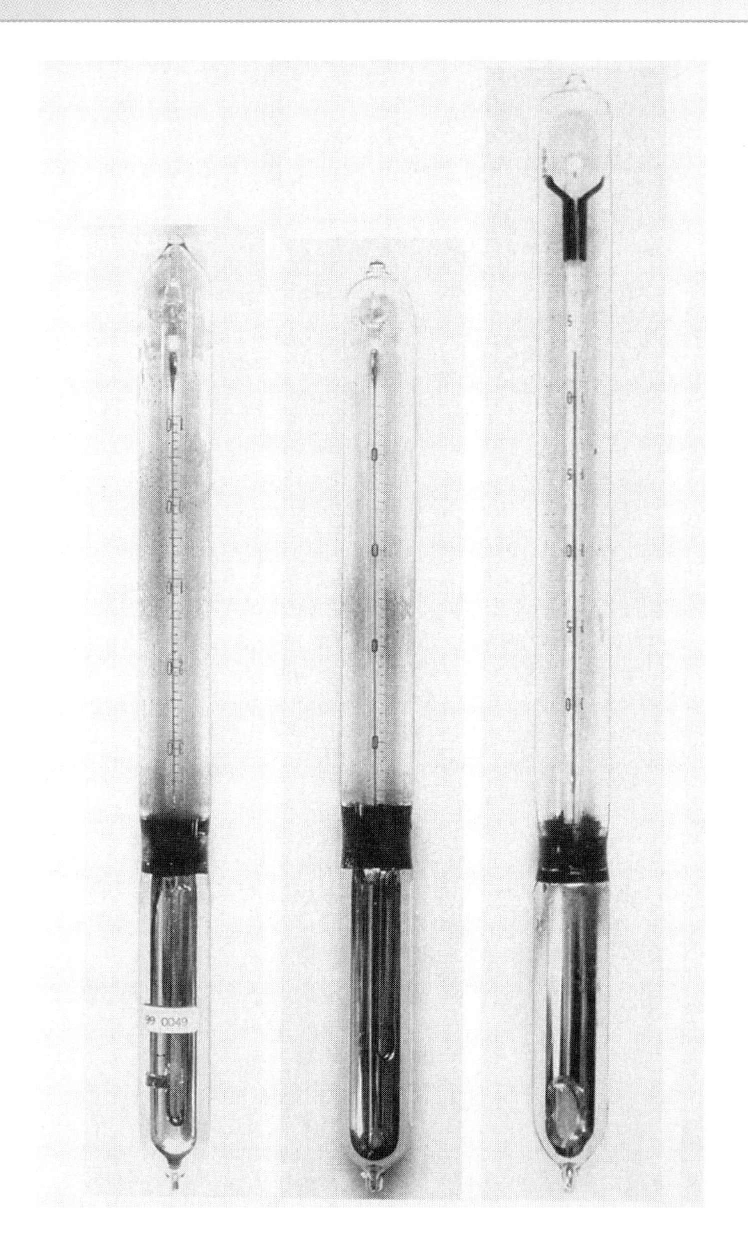

Since mercury is the only metal in a liquid state at room temperature, it has historically been used to make thermometers. These three Chabaud thermometers were constructed by Victor Chabaud in 1892.

Nonmetals, however, are more difficult to pigeonhole. Some are solids in their pure form, some are liquids, and some are gases. Some are essential to all life on Earth, while others are deadly. They are perhaps best defined by contrast: In pure form, they are neither good conductors of electricity nor shiny. And when nonmetals form ions, they tend to take on extra electrons and form negative ions. This latter difference between metals and nonmetals—that metals like to form positive ions, and nonmetals like to form negative ions—means that metals and nonmetals can bond together through electrical attractions, such as the ionic bonds of sodium chloride, which will be discussed in chapter two. Moreover, nonmetals have an ability that metals do not: They can bond together with other nonmetals by sharing electrons, and, as explored in chapter two, form molecules.

Along the staircase division between metals and nonmetals are elements such as silicon and arsenic, known as the semimetals, or metalloids. The metalloids serve to remind us that as the complexity of structure advances, the line that divides the elements blurs. Due to a gradation of properties from metal to nonmetal, the elements at the interface are known as metalloids. The properties of the metalloids are again best defined in relief. The metals are conductors of electricity, the nonmetals are nonconductors, and the metalloids are semiconductors: materials that have changed the face of the world. Anything that is computerized or uses radio waves depends on semiconductors.

There is more to be gleaned from the shape of the periodic table than just the division into metals, nonmetals, and metalloids. Each column of the periodic table represents a related family of compounds, which is why each column is numbered. This column number is called the group number, and elements with the same group number have similar chemical properties.

A Closer Look at the Metals

A chemical family is related by the number of electrons in the outermost shell, the valence electrons. All the elements in the first column, group 1, have one valence electron; all the elements in the second column, group 2, have two valence electrons. Group 18 elements have eight valence electrons, group 17 elements have six valence electrons, and groups 16, 15, 14, and 13 have six, five, four, and three, respectively. At this point the relationship between the number of valence electrons and the group number becomes more convoluted, but any complication that applies to one member of the group applies to the others, so, with a few exceptions, all elements within a group end up having the same number of valence electrons. Because the electrons determine chemical reactivity, having the same number of valence electrons means having roughly the same chemistry.

Above: A gold nugget. Gold is one of the so-called noble metals, which are metals that are resistant to corrosion or oxidation. Top left: Sodium, an alkali metal.

Metals are generally thought of as fairly solid and unyielding, but the metals of group 1, the alkali metals that include sodium, lithium, and potassium, are soft enough at room temperature to be cut with a butter knife. When a freshly cut chunk of an alkali metal is tossed into water, it releases hydrogen and bursts into flame, something not normally expected from a metal.

The group 2 metals, the alkaline earths, are also reactive. Magnesium and calcium, two elements from this group, are essential for life processes. The next series of groups, the elements in group 3 through group 12, are commonly called the transition metals because they form a transition between the metals and the metalloids.

Iron, a transition metal given the symbol Fe for *ferrum* (the Latin word for iron), is commonly considered the stuff of bridges, rockets, and railroads. In truth, these properties may hold for alloys of iron, but if the iron horses that drove the Industrial Revolution had been made of pure iron, the

Platinum is a very heavy precious metal that is used in high-temperature laboratory work where a substance is required that resists chemical attack even when very hot. It is also used in dentistry, for surgical pins, and in expensive jewelry.

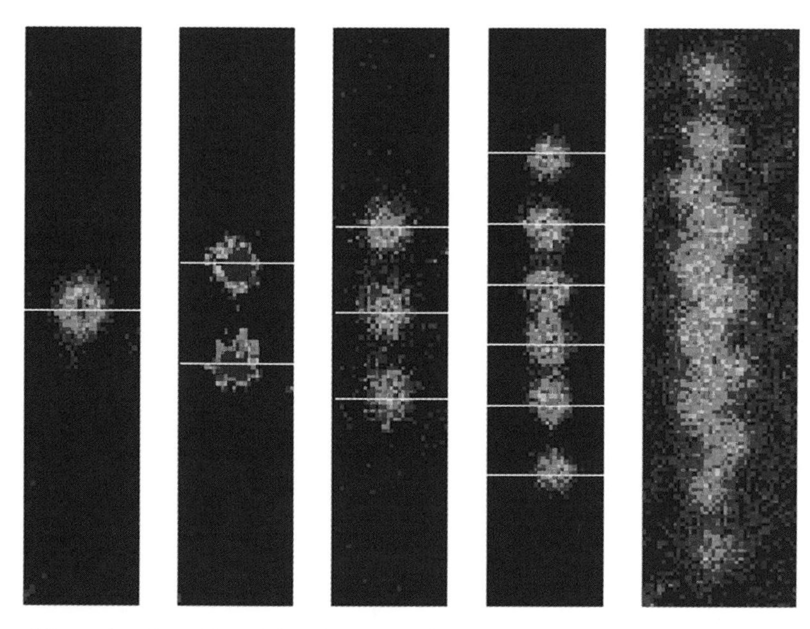

Magnesium ions. These false-color images show 1, 2, 3, 6, and 12 magnesium ions loaded into a machine called an ion trap. Red indicates the centers of the ions.

revolution would have derailed. Iron, its pure form, is a reactive, brittle metal that can rust away in a matter of days.

Nickel, palladium, and platinum, group 10, live up to their reputation for intractability, as do the group 11 metals, copper, silver, and gold. In fact, copper, silver, and gold can be found naturally in their pure, unreacted form. Gold nuggets are panned from Alaskan creek beds, and prehistoric peoples made beads of gathered copper and silver. Other metals must be smelted from chemical compounds of the metal—ores—if the pure form is desired. Copper and gold also have the distinction of being reddish-brown and yellow, respectively, not the lustrous gray of other metals, but silver is, well, silvery. All three metals are used as coinage metals and are sometimes referred to as the noble metals because coins were associated with the aristocracy. Usage, however, has associated the term "noble" with "unreactive" because resistance to corrosion is what made these metals useful for coins. Group 12 consists of zinc, cadmium, and mercury, the only metal liquid at room temperature. Mercury alloys, called amalgams, readily form with many metals, including gold. Mercury was once used to harvest gold from rock.

A group called the inner transition metals is located at the very bottom of the traditional periodic table. Radioactivity is an important property of these fascinating metals, and these materials and their radiochemistry will be explored in chapter six.

Aluminum, tin, and lead are still considered metals because they are under the staircase. These metals are sometimes referred to as the poor metals because they are softer and have lower melting points than the transition metals. Aluminum is so reactive that it is very difficult to separate from its oxide, which is why aluminum, the most abundant metal in the Earth's crust, is still important to recycle.

Aluminum, the most abundant metal in the Earth's crust, is not found free in nature. Refining aluminum from the compound bauxite is more expensive than recycling the metal in its already refined state.

A Closer Look at the Nonmetals and Metalloids

Metals are good electrical conductors. Nonmetals are not. Between these two extremes are metalloids: the semiconductors of the periodic table.

METALLOIDS

Starting at the top of the periodic table, we encounter boron, which is both a micronutrient and poison; boric acid is an effective eliminator of cockroaches. Pure boron is not a semiconductor, but, to produce desired properties, boron can be blended with another metalloid that has a major reputation as a semiconductor, silicon. Silicon is the second-most-abundant element in the Earth's crust, and joined with the most abundant element in the Earth's crust, oxygen, it forms silicon dioxide, also known as silica, the basis of sand, clay, and quartz. To give pure silicon desirable semiconducting properties, silicon is doped; that is, carefully selected and controlled amounts of other metalloids are seeded into the crystal.

Germanium, the next metalloid, was the primary material of the original transistor, though it was superseded when methods for purifying the less expensive silicon became available. Germanium is now used to add silicon because it is in the same group as silicon, so the two have enough properties in common for germanium to slip comfortably into its matrix. One of the famous blanks in Mendeleev's table was neatly filled by the discovery of germanium. Arsenic, a metalloid, achieves its most infamous property—its ability to act as a poison—because it belongs to the same family as phosphorus: It can insinuate itself into positions that require phosphorus in biochemical reactions. Of the remaining metalloids, antimony (given the symbol Sb from the Latin *stibium*) and tellurium have both found application in semiconductor materials. Polonium (named by Marie Curie for her beloved homeland, Poland) is radioactive, as is astitine. Astitine has been dubbed the rarest naturally occurring element. Produced in the radioactive decay of uranium, a fairly common element, it has a lifetime of only about eight hours. Less than a teaspoon of astitine can be found in the Earth's crust at any one time. The elements that do make up the crust of the Earth are more likely to be found higher on the periodic table and a little to the right—the nonmetals.

Above: A cluster of quartz crystals. Top left: Neon signs are created by running an electrical current through neon gas.

Marie Curie and her husband, Pierre, separated the highly radioactive element polonium from uranium ore.

Tellurium is a brittle, silvery-white element that is rarely found in nature as a native element but is often found in other minerals. Tellurium is sometimes added to lead to improve the metal's durability and is also used in ceramics, blasting caps, and semiconductors.

NONMETALS

Although the nonmetals comprise less than a fifth of the known elements, compounds that are formed from these elements number in the tens of millions, and the number is increasing every day. Many of these come from the multitalented element carbon and the many forms of carbon compounds that make up the field of organic chemistry, which will be treated separately in chapter seven. Nitrogen, phosphorus, oxygen, sulfur, and even selenium are also essential to organic life, and their properties will be examined in this context in chapter eight.

Doctors recommend 150 micrograms of iodine in an adult's daily diet. Natural sources of iodine include seaweed and seafood. Table salt is often enriched with iodine and is referred to as iodized salt.

Group 17 is the halogens. The halogen group contains the familiar elements fluorine (which helps maintain healthy teeth), chlorine (the partner of sodium in sodium chloride, table salt), and iodine (lack of which can cause goiters and cretinism). Halogens are generally reactive, which accounts for the activity of chlorine bleach.

Group 18 finishes our survey of the nonmetals—and the periodic table—with a list of an interesting column of elements called the noble gases. These gases possess almost no reactivity. Helium, for instance, is used to fill balloons for children without fear of explosion. We also take advantage of the lack of reactivity when we run a current through neon gas to create neon lights. But under some extreme conditions, xenon (from the Greek word *xeno*, meaning "stranger,") can be changed from an inert gas into a rare and reactive solid. This curious realm of chemical compounds is ripe for examination.

CHAPTER 2

COMPOUNDS

Left: Clouds are made up of condensed water vapor. The hydrogen-oxygen bond in a water molecule is a good example of a polar covalent bond. Top: English chemist and physicist Michael Faraday. Bottom: A rendering of water molecules displaying two hydrogen atoms bonded with an oxygen atom.

When chemists look at the periodic table, they see all the materials that make up the world: the water suspended in clouds, the plastics in baby bottles, the medicines of healers, and the tissue of a beating heart. These materials all come from the coalescence of atoms of the elements on the periodic table. How and why atoms come together to form compounds was a conundrum to scientists in the mid-1800s. Chemists knew at the time that hydrogen and oxygen gases united in a ratio of two to one to form water, H_2O, but they did not know why. When they discussed the bonding together of elements they spoke of a quality called "chemical affinity," but they had no explanation for it.

The English chemist Michael Faraday (1791–1867) had elucidated the attraction of a negative electrical charge for a positive electrical charge in chemical systems. The notion of negative electrons and positive protons in atoms hinted that a theory to explain chemical affinity was at hand. Swedish chemist Svante Arrhenius (1859–1927) showed that charged "species" were able to exist as separate entities in solution, and in the United States, the chemist G. N. Lewis (1875–1946) explained that two electrons were necessary for a stable bond. Once all these ideas merged with the new ideas of quantum mechanics, a comprehensive theory of chemical bonding was born.

Forming Ions

Every atom of a particular element always has the same number of protons, but electrons can come and go. When the number of electrons changes, ions, charged chemical species, are produced. Michael Faraday was able to show that applying a voltage to a chemical solution could produce ions in solution, but when Svante Arrhenius proposed that ions could exist in solution without applied voltage, his ideas were met with derision.

Why should the electrons move around? What determines if an atom will gain or lose electrons and how many? One of the cornerstones of the theory of ionization was laid by G. N. Lewis, an American chemist, who noticed that atoms tended to form ions that had eight electrons in their outer shell. For pedagogical purposes, Lewis presented a simplified version of an atom with electrons forming the corners of cubes; in this way, he showed how the commonly known ions could be predicted by deciding how many electrons needed to be gained or cast off to have a completed outer cube. For instance, chlorine, with seven outer electrons, would want to gain an electron to fill its outer cube. Sodium, with one outer electron, would want to shed this electron to be left with a filled underlying cube.

SHELLS AND SUBSHELLS

Lewis's model fit well with the theory of quantum mechanics under development at the same time because Lewis's cubes corresponded to the shells and subshells. By looking at the periodic table, it is possible to tell how far from a filled subshell a particular atom is by seeing how far it is from the end of a row. Sodium (Na), which is in the first column of the periodic table, will tend to lose its lone electron to establish a filled shell, and chlorine (Cl), in the seventeenth column, will tend to gain an electron to establish a filled shell. In the compound sodium chloride, sodium's need to shed an electron and chlorine's need to gain an electron mesh to cause the combined elements to form a stable arrangement.

BONDING MECHANISMS

Lewis's model also provided a rationalization for another important concept: the two-electron bond. Lewis showed how the joining of two cubes could be accomplished by joining a common edge, and an edge comprises two corners—two electrons.

Lewis's model was not meant to be taken literally. Earlier in the century, German theorist Werner Heisenberg (1901–76) showed that it was theoretically

Above: Swedish chemist Svante Arrhenius received the Nobel Prize for Chemistry in 1903. Top left: Salt crystals. The compound sodium chloride is held together by an ionic bond.

impossible to know the position and momentum of an atomic-sized particle at the same time. To measure an atom's position, light would have to be reflected from the atom. But light is sufficiently energetic to knock an atom off course, so the measurement would change the momentum of the atom. A caveat of this principle is that an atomic-sized particle can never be completely still. If it were, the position and momentum, which would be zero, would be simultaneously known.

So although electrons cannot literally be located at the corner of cubes—the Heisenberg principle forbids electrons from remaining stationary—Lewis's notion that a pair of electrons are required for a bond became a major tenet of the quantum mechanical model of chemical bonding. The mechanism by which this joining is accomplished, however, can vary. There are situations best explained by an electron transfer from one element to another to form ions and an ionic bond, such as the table salt discussed above, and there are situations where sharing electrons is favorable and what is known as a covalent bond is formed.

Werner Heisenberg, whose work in physics impacted G.N. Lewis's model of chemical bonding.

The cities of Chicago, Illinois, and Jersey City, New Jersey, started using water chlorination techniques in 1908. Chlorination plants such as this one in Minneapolis, Minnesota (c. 1911) helped to combat such diseases as cholera, typhoid fever, dysentery, and hepatitis A, saving thousands of lives.

Ionic Compounds

To a chemist, table salt is not the only salt. Ionic compounds, compounds formed from ions, are also commonly called salts. For instance, magnesium sulfate is known as Epsom salts, and potassium nitrate is known as saltpeter. So-called "lite" salt, used by people trying to reduce their sodium intake, is potassium chloride. Virtually all ionic compounds are soluble in water to some extent, though some are very soluble, such as table salt, and some are fairly insoluble, such as baking soda. Salts can be added to solutions of ions to crowd other salts out of solution. This process, called salting out, is used in the soapmaking process.

ANIONS AND CATIONS

But what type of ion is an atom of an element likely to form? The observation that an atom of an element will then tend to gain or lose the number of electrons it needs to have a filled shell—and do this in the least number of moves—means that the elements on the left side of the periodic table will be much more likely to form positive ions than negative ions. Elements on the left need to lose only one or two electrons to gain a filled shell. Elements on the right side of the periodic table are much

Common table salt is highly soluble, meaning it is easily dissolved into another substance.

A giant soap kettle in an industrial plant, c. 1942.
Top left: Essentially all ionic compounds are soluble in water.

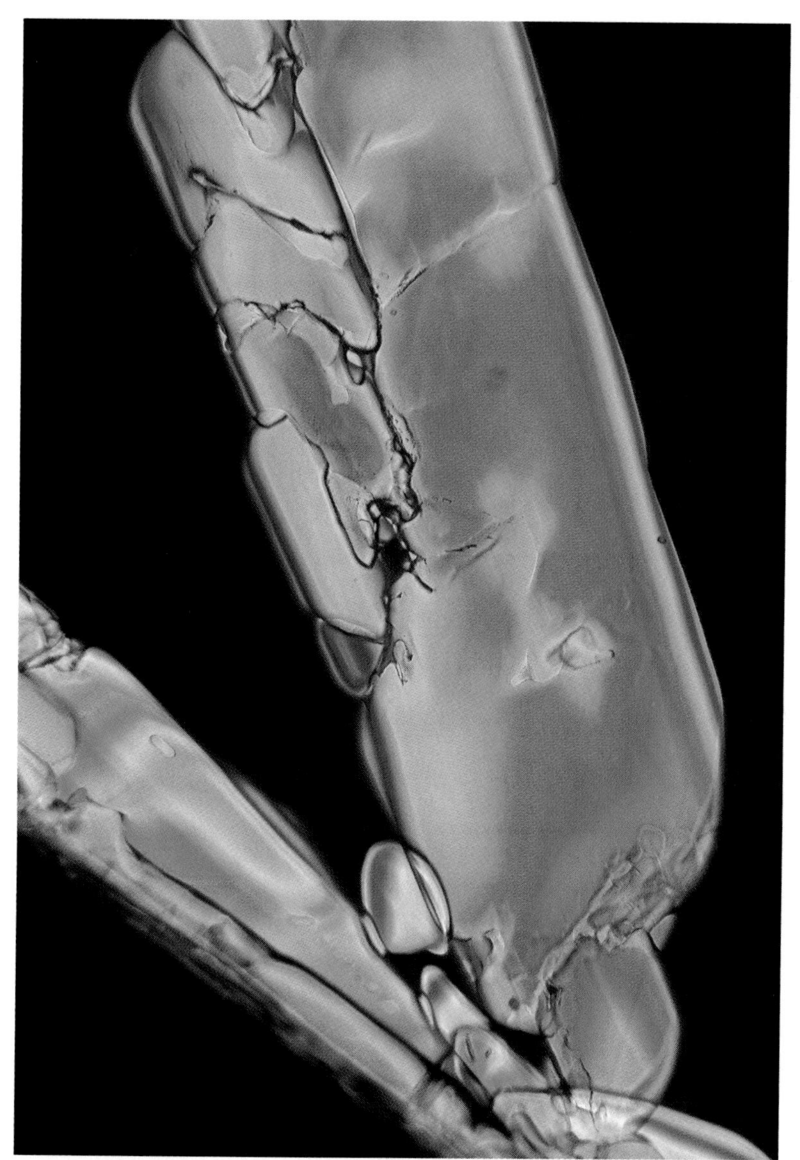

Magnesium sulfate crystals. This compound in its hydrated form is known as Epsom salts, a bitter-tasting substance used medicinally as a purgative and in hypertonic baths to soothe the skin and ease muscle tension.

Making soap over an open fire in Henryville, Pennsylvania, around 1900.

FRONTIER SOAP-MAKING

A tricky, messy, and sometimes downright dangerous job, soap-making was nonetheless necessary for the minimum creature comforts in the rugged conditions of the North American frontier. Although the techniques were handed down as family recipes and the soapmakers obtained their skills by participation, the process was nothing less than chemistry and the soapmakers, mostly women, were, for all practical purposes, pioneer chemists.

The first step in the process was to make caustic lye, a strongly corrosive solution of sodium hydroxide ($NaOH$), by running water through wood ash. Animal lard from butchering and saved table scraps was boiled with the lye until the mixture was thick. Salt was then added to force the soap out of solution and into a hard paste that could be cut into cakes. After all this work, not a scrap of soap was ever discarded or wasted!

more likely to form negative ions because they only need to gain a few electrons to fill their shells. Therefore, elements to the left tend to form positive ions, *cations*, and elements to the right will tend to form negative ions, or *anions*. Cations are electrically attracted to the anions, and when they meet and marry, the result is an ionic bond, which results in an ionic compound. But what about compounds formed from elements that are on the same side? Who wins the tug of war for electrons? The answer is that these elements have found a peaceful solution: When elements have a similar attraction for electrons, they can form a chemical bond by sharing. When valence electrons are shared to fill subshells, then the bond is called covalent.

Covalent Compounds

Nonmetals seek each other out to form covalent bonds. Why? To be mutually fulfilling—of octets, that is.
By consulting the periodic table we can see that chlorine (Cl), has a filled first shell and a filled second shell, but only seven electrons—one shy of a full octet—in its third shell. But if two chlorine atoms join forces, if they share one electron each, then there will be eight electrons available to fill the octet of both chlorine nuclei and everybody is happy. Hydrogen has to find an electron to share to fill its first outer shell with two electrons. However, luckily for all forms of life on Earth, it does. Two hydrogen atoms are able to share their lone electrons with the six electrons of oxygen so that oxygen ends up with a filled shell of eight and each hydrogen atom ends up with a filled first shell of two electrons, and we end up with the useful material H_2O, water.

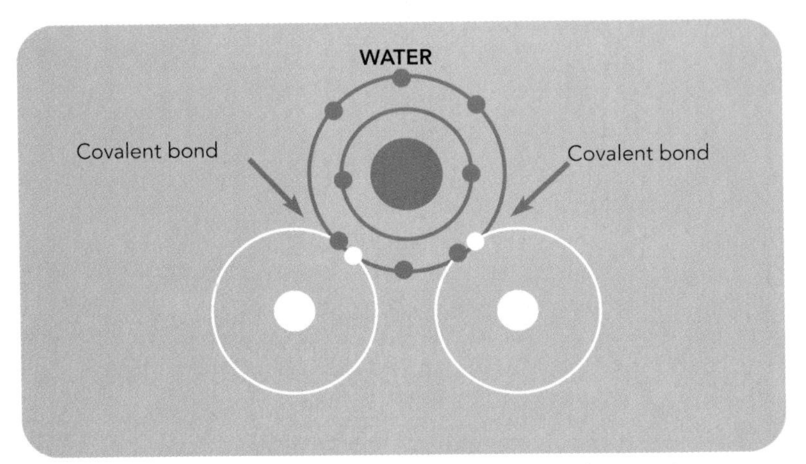

This diagram shows the covalent bonds between hydrogen and oxygen that are found in water. Top left: Acetylyne, which is a compound of carbon and hydrogen (C_2H_2), is the fuel used in welding torches.

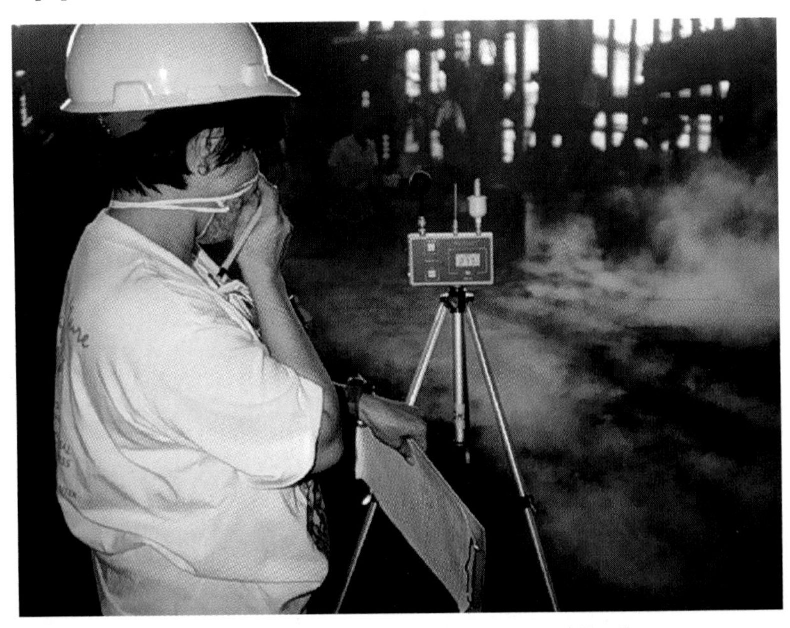

An inspector with the Philippines Department of Labor and Employment measures carbon monoxide (CO) levels, as well as the number of silica and metal particulates in airborne dust. High levels of carbon monoxide can kill a person in minutes.

ELECTRON SHARING

At times, sharing just one pair of electrons is not enough. To ensure that everyone has a happily filled octet, sometimes two or three pairs are shared. In carbon dioxide, CO_2, the central carbon shares two pairs of electrons with each oxygen in an arrangement called a double bond. Acetylene, used as a fuel in welding torches, has the formula C_2H_2 and is formed by sharing three set of electrons between the carbons in a triple bond. This conviviality on the part of nonmetals means that some groups of elements have several stable bonding arrangements. For instance, carbon and oxygen can form carbon dioxide, CO_2, or carbon monoxide, CO. Nitrogen and oxygen can combine as NO_2, N_2O_4, N_2O_5, N_2O, or NO.

To the number of compounds formed from covalent bonds, add the number of compounds formed with ionic bonds, and the need for an organized system of nomenclature becomes apparent—a system that accommodates all possible compounds without allowing for duplication. The system that evolved, as we shall see, accomplishes just that.

Liquid nitrogen being added to a thermally insulated container. Liquid nitrogen is created industrially by distilling liquid air, or air that has been liquefied by compression.

American chemist Gilbert Newton Lewis in 1916.

WORKING FOR CHANGE: G. N. LEWIS

Gilbert Newton Lewis learned to read at the age of 3 and was taught at home by his parents until he was 9 years old. He received his Ph.D. in chemistry from Harvard university when he was 24 and was a full professor at the Massachusetts Institute of Technology (MIT) by the time he was 36. He left MIT for the University of California at Berkeley the next year.

Lewis was a physical chemist, chemists whose interest lie in the theory of chemicals and chemical reactions. He published papers on an enormous range of topics: relativity, magnetic properties of materials, phosphorescence, and the nature of the chemical bonds, among others. In addition to being a creative and insightful theoretician, he was an able experimentalist. He died, at the age of 70, in his lab.

Names, Symbols, and Formulas

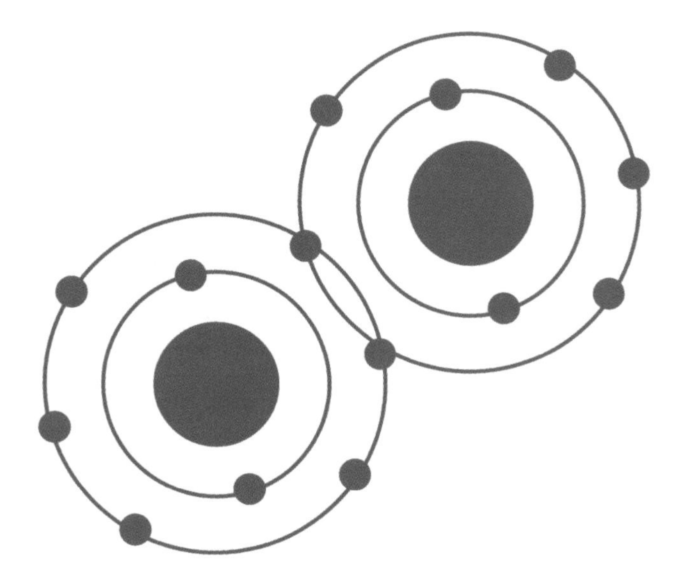

Above: Two oxygen atoms bonding. An oxygen molecule's simple formula is O_2. Top left: Many household cleaners include the ammonia ion called ammonium, which is represented as NH_4.

Though H_2O stands for that wonderfully versatile and useful material, water, it is extremely simple as far as chemical formulas go. Other molecules, such as oxygen (O_2) have simple formulas, too, but some, such as insulin ($C_{257}H_{383}N_{65}O_{77}S_6$), do not. Chemical nomenclature, or the systematic naming of chemical compounds, is based on two guiding principles: No two different materials can share the same name, and each unique name must be the simplest name possible. To satisfy these requirements, organic compounds, covalently bonded materials formed from carbon and hydrogen, are named separately from inorganic compounds. Systemized organic nomenclature will be discussed in chapter 7, after a little of the nature of organic chemistry has been explored. Inorganic compounds are used here to demonstrate the principles behind chemical nomenclature.

For nomenclature purposes, inorganic compounds are further subdivided into ionic and covalent compounds. In an ionic compound, the positive ion, the cation, is named first and given the name of the metal from which it was formed. The negative ion is named second and given the suffix "ide" to indicate that it is an ion; for example, sodium chloride (NaCl).

COVALENT COMPLEXITIES

For covalent compounds, however, the situation can be more complex because often the same two elements can form several covalent compounds. For instance, carbon and oxygen can form carbon monoxide (CO) and carbon dioxide (CO_2). The solution to the problem is also exemplified by these two compounds: The number of atoms of a particular element that goes into a covalent compound is indicated by the prefixes mono, di, tri, tetra, and so on. In the interest of simplicity (one of the primary dictums of our nomenclature system), the "mono" prefix is omitted if it can be inferred.

Certain groups are often considered as one unit, such as the hydroxide ion, OH^-, in sodium hydroxide, or lye. When there are two or more of these groups in a compound, they are placed in parentheses and a subscript outside the parentheses indicates the number of groups. Thus a material often used by gardeners to adjust the alkalinity of the soil, $Ca(OH)_2$, commonly called slaked lime, would be named systematically

Exhaust from gasoline-powered automobiles contains a number of chemical compounds, including carbon monoxide (CO), carbon dioxide (CO_2), and sulfur dioxide (SO_2), which stimulates acid rain.

as calcium hydroxide. Common ion groups, or polyatomic ions, that are treated as single units include the ammonium ion, NH_4^+; the nitrate ion, NO_3^-; and the bicarbonate ion, HCO_3^-. The ammonium ion is present in many household cleaning solutions; nitrates have been used as food preservatives, in gunpowder, and as another soil nutrient; and the bicarbonate ion is familiarly found in sodium bicarbonate: baking soda.

There are other situations not covered in this brief survey, but perhaps this overview will provide an idea of how chemical nomenclature is accomplished. The formulas of the compounds have been named as a matter of observed fact, but the question remains: Why do bonds form in the first place?

A worker fertilizes a tea field. Nitrates, which are inorganic compounds, are often used as fertilizer.

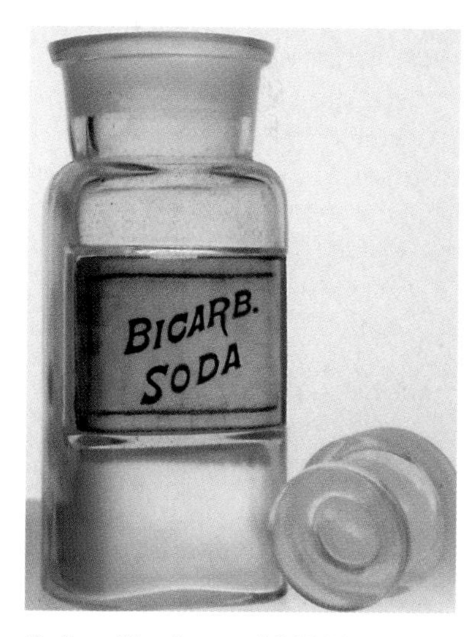

Sodium bicarbonate ($NaHCO_3$), commonly known as baking soda, is a staple in baking, a fire extinguishing substance, and an antidote for indigestion.

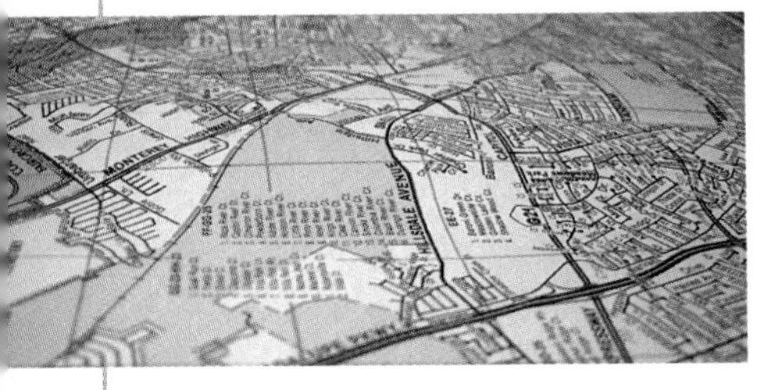

Electrons and Chemistry

In ionic bonds, the attraction between opposite charges pulls the ions together. Covalent bonding results from sharing electron pairs, but this model introduces a troubling question: If negatively charged electrons repel, how can pairs of electrons be shared in a covalent bond?

THE TWO-ELECTRON BOND

As the electron cloud of an approaching atom feels a pull from the oppositely charged nucleus of another atom, the atoms can be drawn together. At the proper distance, the negative electron pairs arrange themselves between the two nuclei so that they maximize their alignment with the positive nuclei—and minimize the repulsion felt by the two nuclei—creating a bond. The distance at which attractive and repulsive forces are balanced is called the bond length.

But if charge–charge repulsion is important in determining bond length, why don't repulsions between the negatively charged electrons disrupt the bond? It turns out there is another consideration that allows the electrons to be at least somewhat together. This second consideration is something called electron spin.

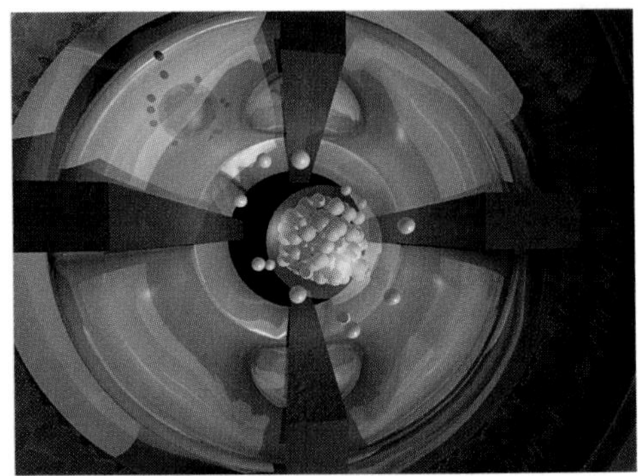

This illustration shows a single atom of the element americium-243 at the center of a cyclotron particle accelerator. The nucleus of protons and neutrons is surrounded by an electron cloud. Americium is an artificially produced metal.

Above: Bales of sorted recyclables are stacked for processing at the Rumpke Recycling Center in Columbus, Ohio. Although the facility is losing money on its recycling business, it continues the service to save room in limited landfill space. Top left: In much the same way that maps are used to describe the structure of a city, molecular and subatomic models are used to illustrate the behavior of molecules and electrons.

ELECTRON SPIN

It must be established from the onset that "electron spin" does not mean that the electrons are actually spinning. A spinning electron would be a moving electrical charge, so it would generate a magnetic field, as an electromagnet at a recycling center uses electricity to generate the magnetic force necessary to sift metals. The electron does have a measurable magnetic moment, but if this magnetic moment were produced by spinning, then the electron would have to be spinning faster than the speed of light, which is forbidden by physics.

The concept of spin, however, is useful. If one electron

were spinning in one direction and another electron were spinning in the other direction, then the magnetic fields of the two electrons would be opposite and they would attract each other. Two electrons thus paired could share a closer space, such as the region between nuclei, and form a bond.

VSEPR

It cannot be too strongly stated that these models are just that—models—in the same sense that a map is a model of a city. All the detail of the city cannot be included in the map because that would render the map useless. All the aspects of electrons and their behavior cannot be covered by one macroscopic model. The various models do, however, have their usefulness, and this usefulness does extend to real-world consideration. For instance, the model of two electrons per bond can be used to predict the important property of molecular three-dimensional shape, a concern of considerable consequence for the behavior of bulk materials. The theory that relates the electron occupancy of bonds and nonbonding orbitals to three-dimensional shape is called the Valence Shell Electron Pair Repulsion, VSEPR. According to the tenets of VSEPR, the electron pairs in bonds are content to be with each other, but there is repulsion from pair to pair. Like hair on a dry day, the electrons spread out to achieve the maximum distance between pairs. This repulsion is responsible for the three-dimensional shape of molecules, which in turn is responsible for the three-dimensional shape of the materials that make up our world.

These models—the octet, the two-electron bond, VSEPR—that explain many of the behaviors of materials are a tribute to the theoreticians and experimentalists who have made chemistry their study. So much progress is remarkable considering what we are working with—atoms and electrons in a constant state of motion with electric and magnetic interactions—and the fact that these particles are all so darn small.

A molecular model of iodine pentafluoride, a toxic gas. Iodine is shown in orange, fluorine in red. In blue is a pair of electrons not involved in the chemical bond.

A smelter at a mining company uses an electromagnet to remove metallic particles from ore. Electromagnets use electricity to generate magnetic force. Electrons, nature's smallest magnets, are believed to have an intrinsic magnetic field.

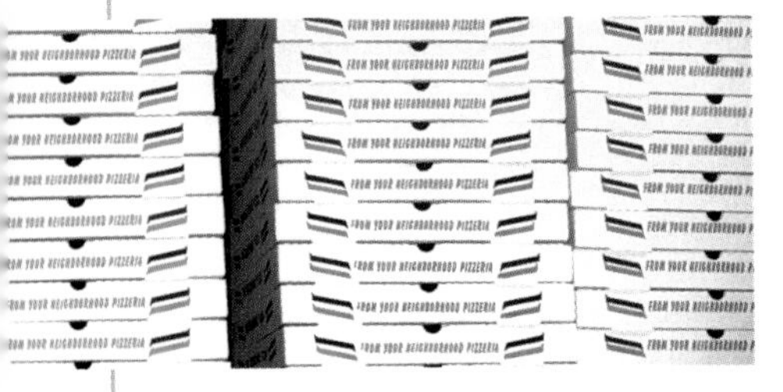

A Matter of Scale

The actual size of atoms is truly difficult to imagine. In fact, at the dawn of the twentieth century, there were still those who questioned the legitimacy of the atomic model because there was no definitive proof of atoms' existence. The evidence, when it did come, was from a rather surprising quarter.

Robert Brown (1773–1858), a botanist, observed that tiny particles such as dust and pollen bob around in liquid as though they have a life of their own. He detected this "Brownian" motion at all times of day and in all locations. In the early 1900s, the physics community, including Albert Einstein (1879–1955), had interest in the kinetic molecular theory, a theory that assumed gases consisted of particles in constant, random motion and that this motion increased with temperature. Using the kinetic molecular theory, Einstein derived equations that showed how far a gas particle should move in a given amount of time, at a given pressure and temperature, assuming that it had been buffeted about by the random motion of particles in a gas, but he did not know how or if the effect could ever be measured.

Above: Albert Einstein. Top left: In the International System of Units, a mole describes an amount of a substance. One mole of dollars will buy you 17 billion stacks of pizza.

Jean Baptiste Perrin developed Avogadro's number, which is the number of particles in a mole.

BROWNIAN MOTION

Then came the French physical chemist Jean-Baptiste Perrin (1870–1942). Perrin recognized that Robert Brown's random motions of pollen grains were the result of buffeting by atomic-scale particles of liquid.

The motion of Brownian particles can be modeled by a sock on laundry day. If the laundry ready for washing is piled on the floor of a house full of rambunctious children, the sock may suffer random kicks as the children run through the house. Sometimes the kicks will be in one direction, and sometimes in another, but at the end of the day, the sock will have moved away from the main pile. The distance that the sock moves can depend, statistically, on how many children are running in the house. When Jean Perrin measured the movement of his Brownian particles, he was able to use Einstein's equations to calculate the number of particles in a given volume of liquid. Perrin called the number he measured "Avogadro's number," in honor of Count Amedeo Avogadro (1776–1856), another innovative theoretician in the history of the atomic

One-mole samples of several chemical substances. A single mole of any substance contains the same number of molecules or atoms, represented by Avogadro's number. The weight of each molar sample is determined by the weight of the molecules and is calculated from the weights of the atoms in the molecule.

theory. Today we use Avogadro's number as the standard count for atomic-sized particles. It is the number of particles in a convenient amount known as a mole.

A drop of water the size of the period at the end of this sentence would contain 10 trillion water molecules. Considering numbers of this magnitude is daunting, so scientists developed a calculation for quantity called the mole. This allows chemists to better manage their study of the elements in the world around us.

A WORLD OF MOLES

A mole is a certain number of items, just as a dozen is twelve items. A dozen is a convenient number of eggs or doughnuts, but a dozen atoms would be too little to weigh, measure, or even see. To have a reasonable number of atoms to work with, a much larger number is needed, and the number chemists settled on was a mole. But while a dozen is 12 things, a mole is 602,213,673,600,000,000,000,000 things—which is just a few more. This number might be more understandably expressed as 6.022×10^{23}.

A dozen eggs fit neatly in a refrigerator door, but a mole of eggs would cover the entire surface of the Earth with stacks of eggs taller than three hundred Empire State Buildings. A dozen dollars might buy one pizza, but a mole of dollars would buy enough pizzas to make about 17 billion stacks of pizza to the Sun!

Why such a ridiculously large number? Because atoms are so tiny. A mole of carbon atoms is only about a handful. A couple of moles of helium is about a bathtub full. But down at the atomic level, these tiny atoms mix and combine into the vital substances and mixtures that make up our mole-sized world.

Substances and Mixtures

Classification schemes are always useful, such as the familiar classification of living things into kingdom, phylum, class, order, family, genus, and species. Chemists, it turns out, classify, too.

The first categorization for chemists is the division of the world into energy and matter, the famous *E* and *m* of Einstein's equation, $E = mc^2$. Although the interpretation of Einstein's equation is that matter and energy can be interchanged, the saving grace for chemists is the *c* in Einstein's equation—which stands for the speed of light. Chemists are hard workers, but they don't work at the speed of light, so for the purposes of chemistry, energy and matter can be considered separate and immutable. In the world of benchtop chemistry, the principles of the conservation of matter and the conservation of energy are firmly in place.

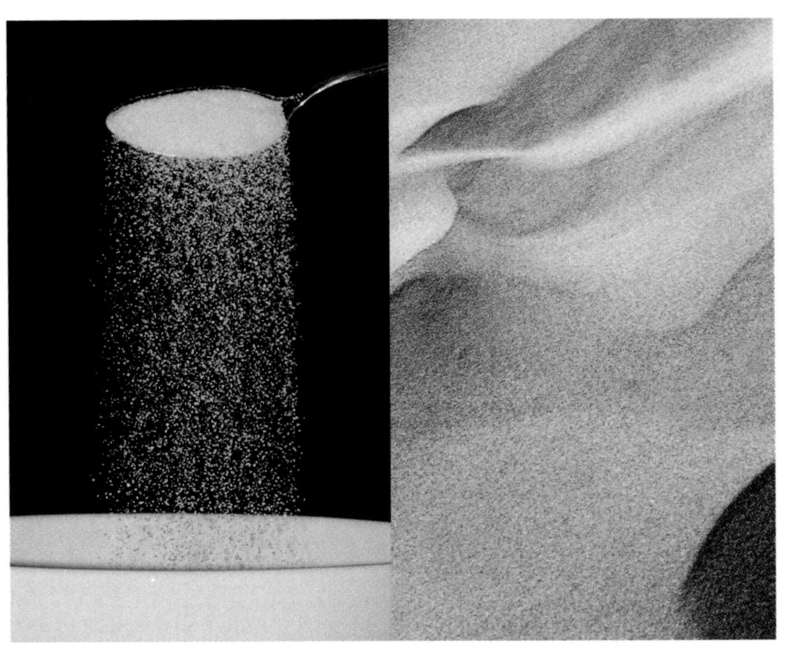

Mixtures such as sugar water and sand have variable compositions. Even if the ratio of sand's component parts is altered, the mixture is still sand. Top left: Water is a pure substance, which means that the ratio of its parts—two hydrogens to one oxygen—cannot be varied.

MIXTURES AND PURE SUBSTANCES

Of the two, energy and matter, matter can further be subdivided into mixtures and pure substances. The difference between mixtures and pure substances is variability. The ratio of parts in a pure substance, say in H_2O, cannot be varied without changing the nature of the substance. Water has a ratio of two hydrogens to one oxygen, and if this is changed to two hydrogens to every *two* oxygens, a new substance results, H_2O_2, hydrogen peroxide. Mixtures, on the other hand, can have variable composition. Sand is a mixture of silicates, carbonates, phosphates, and other minerals, but the composition can vary from black sand to white; it is still all sand. Sugar water can be made in any proportion from sweet water to syrup, the ratio of sugar to water being virtually infinitely variable.

Mixtures such as sand and sugar water can be further subdivided into heterogeneous and homogeneous mixtures. Homogeneous mixtures are uniform throughout, such as sugar water. A sample from the bottom of a glass of sugar water should have the same composition as a sample from the top. The situation is different for sand. The composition of sand

Top: The element copper (Cu) has 29 protons. It is an excellent conductor of heat and is therefore favored for use in cookware. Bottom: Copper sulfate is a chemical compound ($CuSO_4$) commonly used as a fungicide or herbicide; these workers are applying copper sulfate to kill parasite-harboring snails.

varies widely from spot to spot, and the sifting of sand from the natural movements of Earth account for the concentration of lighter elements in the Earth's crust, which makes Earth's soil a heterogeneous mixture.

Pure substances, likewise, can be subdivided into compounds and elements. The elements are the separate pure substances on the periodic table. Each element is composed of like atoms, each with the requisite number of protons. Compounds are materials that are formed from specific combinations of elements. Copper (Cu) is an element, but copper sulfate (root killer) is a compound.

CHEMICAL VS. PHYSICAL

Classification helps explain another important concept in chemistry: the difference between physical and chemical processes. When hydrogen peroxide, H_2O_2, a disinfectant, decomposes into hydrogen, H_2, and oxygen, O_2, this is a chemical change. The substances produced are a gas that could be explosive under other circumstances (hydrogen) and a gas necessary for life (oxygen). When a pure substance is extracted from a mixture, this is a physical change because it is accomplished without changing the chemical nature of any of the substances involved. For instance, sugar can be recovered from sugar water by allowing the water to evaporate, so evaporation is a physical change. The next chapter looks at another type of physical change: the change from solid to liquid to gas.

909

CHAPTER 3

MATTER IN ALL ITS FORMS

Left: Water exists in three states, or phases. The gaseous state of water is steam. Top: Water in its liquid phase covers over 70 percent of the Earth. Bottom: Sea ice on McMurdo Sound in the Antarctic.

It may seem a bit surprising that all the wonderful complexity and variety of the world—unique leaves and snowflakes; virtually infinite combinations of DNA; billions of species of living organisms—should be possible from such systematically structured atoms. But when we consider the more than 100 different elements currently known and the number of compounds that can be built from combinations of their atoms, the origin of this complexity becomes more understandable.

The variations on materials do not end there. Materials can also be found in different phases: gases, liquids, and solids. Water, a familiar friend, is welcomed in our lives daily in the form of water vapor, liquid water, and ice, the solid form of water. In mixtures, different materials can combine in different ratios, making solutions as varied as saline solution that can soothe the eyes or a pickling solution that can shrivel cabbage to sauerkraut. Solids can mix with solids in alloys, gases with gases in the air we breathe, and solids, liquids, and gases can mix in solutions as different as syrup, vodka, and soda. The permutations and combinations of materials in different phases and infinite ratios conspire to create the soup of the cosmos, and all the variety of animate and inanimate matter.

Solids and Liquids

The characteristics of the solid phase can be understood by contemplating a salt crystal. The basic unit of salt consists of one ion of sodium and one ion of chlorine joined together by an ionic bond. The positively charged sodium ion, however, can attract another negatively charged chloride ion, which attracts another sodium ion, and so on, until ion-to-ion links build a crystal with equal sides and perfect right angles, the uniformity of which is evident even when quadrillions of units are combined to form the single grain of salt. Even to the naked eye, the sharp corners of the crystal are evident, and under a magnifying glass or a microscope, they become as elegant as the facets of a finely cut gem.

All this rigidity of form, however, may lead to the misconception that the atoms

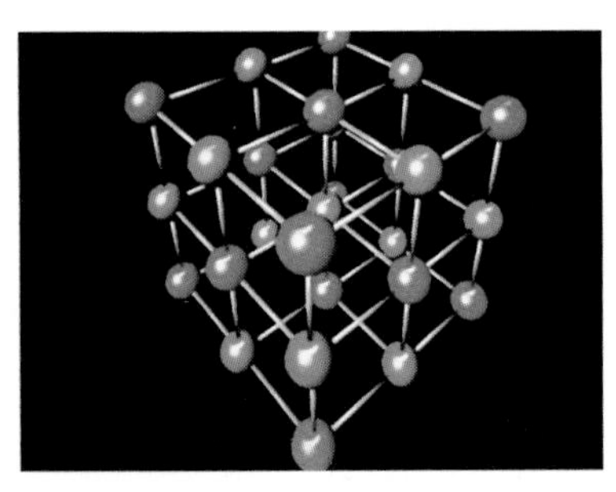

This illustration shows the cubic crystal lattice of sodium chloride. Sodium ions are represented by red spheres, chlorine ions as green spheres.
Top left: A sea salt crystal.

A miner picks gold out of a pan. The process of panning can be used to illustrate the chemical process of annealing, in which the motion of atoms in a solid is increased to allow them to settle into their lowest energy positions.

Olive oil and balsamic vinegar separating. Metals behave in much the same way during the process known as alloy aging.

in the solid are held rigidly in place and exhibit no motion whatsoever. This is not true. The kinetic molecular theory, an elegant model that found its beginnings in the energetic Industrial Revolution, explains intricacies of energy transfer, the behavior of gases, rates of chemical reactions, and many more phenomena based on the assumption of constant atomic and molecular motion. So atoms in solids do indeed experience motion, from vibrating in their crystalline position to exchanging places with other atoms.

ANNEALING

Alloy aging can occur when the mixture of metals that make up the alloy gradually move apart or pool together, in the manner that an oil-and-vinegar salad dressing might settle out. The process in a solid alloy, of course, would be much slower, but it could be accelerated by the application of heat; this is known as annealing. Annealing is any process that uses heat to increase the motion of atoms or groups of atoms within a solid so that they can settle into their lowest energy positions. Annealing might be likened to panning for gold: the shaking of the sand in a watery solution can cause the heavier particles of gold to settle to the bottom of the pan, the lowest energy position. If the application of heat is extreme, a metal might eventually melt, that is, become a liquid. In this case, not only do the nuclei move with respect to each other, but they become so mobile that they can change the shape of the entire sample. In fact, this is the property that distinguishes between solids and liquids: A liquid assumes the shape of its container, a solid does not. With the application of enough heat, pure substances and many compounds will enter a third phase—where the material not only assumes the shape, it also takes up the entire volume of the container: the gas phase.

Atoms in a liquid move around freely within a fixed volume. Although liquids conform to the shape of their containers, the constituent atoms do not form the kinds of rigid structures characteristic of solids.

Gases and the Gas Laws

Gases are characterized by pressure, volume, temperature, and amount of gas, usually measured in moles. These characteristics are interrelated and vary in a predictable fashion known as the gas laws. Consider first a system with a fixed temperature and a fixed amount of gas; for instance, a sample of air trapped in a capped-off syringe. If the pressure on the sample increases, which could be accomplished by pushing the plunger of the syringe down, the volume decreases. When the pressure is released, the plunger springs back, and the volume increases. This relationship is known as Boyle's law, after Robert Boyle (1627–91), one of the early scientists to describe the interdependence of pressure and volume.

Above: Instruments used to measure atmospheric conditions. The balloon is equipped to measure variables, such as pressure, temperature, and humidity, from the air. Other instruments in the background measure temperature, relative humidity, pressure, and precipitation on the ground. Top left: Illustration of the landing of the first hydrogen balloon flight from Paris on December 1, 1783. Experiments with hydrogen balloons at this time led to deeper understanding of the nature of gases.

Robert Boyle. Boyle's law states that under conditions of constant temperature and quantity, there is an inverse relationship between the volume and pressure for an ideal gas.

CHARLES'S LAW AND AVOGADRO'S LAW

The pressure of the atmosphere—the sea of gases we live in, breathe from, and move through every day—may vary significantly in inclement weather. Dropping mercury levels in a barometer, the instrument that measures air pressure, have long been understood to be a portent of foul weather. But on a fair day, the ambient pressure remains virtually constant, so we can consider an empty, capped, plastic soft-drink bottle to be a fair representation of a system with a fixed amount of gas at a fixed pressure. If an empty soft-drink bottle is rinsed with hot water before recycling and immediately tightly capped, the bottle can sometimes be seen to collapse in on itself. If a bottle is placed in the freezer after rinsing with hot water and quickly replacing the cap, the bottle will implode so quickly that audible snaps and crackles can be heard. The reason for the bottle's collapse to a smaller volume is that the volume occupied by cool air is much less than the volume occupied by hot air, an example of Charles's law, named after Jacques Charles (1746–1823).

The fact that volume increases directly with the amount of gas at constant temperature and pressure can be demonstrated by blowing up a balloon. Assuming the temperature and pressure in the room remain constant while you are blowing, the only change causing the volume of the balloon to increase is the amount of air you blow in. By assuming gas samples at identical temperatures and pressures contained the same number of particles, Amedeo Avogadro solved the molecular mystery of oxygen gas—showing that it is a diatomic; that is, molecules made of two atoms—which provided the definitive evidence for discrete molecules to match Dalton's discrete atoms. In Avogadro's honor, this relationship between the volume of a gas and the amount of a gas is named Avogadro's law. It was also after Amedeo Avogadro that Jean Perrin chose to name the number of particles in a mole: Avogadro's number.

IDEAL GAS LAW

When deriving Boyle's law, Charles's law, and Avogadro's law, only two of the gas parameters are allowed to vary and the other two are held constant. While this approach aids understanding, in most practical situations involving the gas phase, temperature, pressure, volume, and the amount of gas are all apt to be changing at the same time. Though it is possible to connect all four factors in one equation, known as the ideal gas law, there are still problems when it comes to some practical applications such as weather prediction. At normal terrestrial temperatures and pressures, the gases of the atmosphere are moving so slowly and are so close together that the result is a material that can have fluid properties as well as gas properties. The ideal gas model becomes much better when the gas sample is heated or when the gas is at very low pressure. But while a pressure can be lowered virtually infinitely, there is a limit to how hot you can heat a gas. If you heat a gas *too* hot, then it will eventually explode!

In a room with constant temperature and pressure, the only thing causing a balloon's size to increase is the amount of air being forced into it. Ideal gases are characterized by three absolutes: pressure, volume, and temperature.

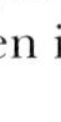

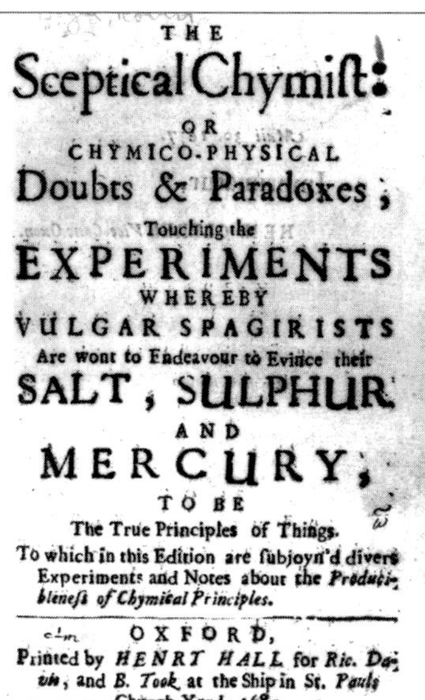

THE

Sceptical Chymiſt:

OR

CHYMICO-PHYSICAL

Doubts & Paradoxes,

Touching the

EXPERIMENTS

WHEREBY

VULGAR SPAGIRISTS

Are wont to Endeavour to Evince their

SALT, SULPHUR

AND

MERCURY,

TO BE

The True Principles of Things.

To which in this Edition are ſubjoyn'd divers Experiments and Notes about the *Producibleneſs of Chymical Principles.*

OXFORD,

Printed by *HENRY HALL* for *Ric. Davis*, and *B. Took* at the Ship in St. *Pauls*

Top: French chemist, physicist, and aeronaut Jacques Charles. Charles's law indicates that for a fixed amount of gas at a fixed pressure, the volume is proportional to the temperature. Bottom: The title page from Robert Boyle's work titled The Sceptical Chymist.

Plasma—the Most Common State

If a gas is heated hot enough, the collisions between the atoms can be so energetic that the gas can be ionized; that is, the electrons can be stripped from the nuclei, leaving a soup of highly energetic charged particles called plasma. Plasmas are created in rocket exhausts, Bunsen burners, and welder's arcs, the latter of which highlight a unique property of plasmas: the ability to conduct electricity. Plasmas created by high voltage have been used to create light in neon signs, sodium vapor lamps, and fluorescent light fixtures—which highlights a second unique feature of plasmas: they glow. Collisions and ionization can cause electrons to find themselves is orbitals above the lowest energy state. When they fall back into a lower energy orbital, energy is released in the form of light.

A plasma arc welder. The torch used in this type of welding contains an electrode that forces the plasma through a fine-bored nozzle, constricting the arc. The plasma exits the instrument at high velocity and temperature, making this technology preferable to other welding processes.

The fourth state of matter was identified by English physicist Sir William Crookes in 1879. Since that time, plasmas have pushed technology forward and are used in television screens (above) and neon lights (top left).

A flat-screen plasma display for a television is composed of a grid of minuscule cells that contain a gas such as neon or xenon. Electrodes apply a high voltage to a cell when its color is needed, and the high voltage causes the gas to transition to the plasma state. The plasma state emits light, which in turn excites colored phosphors to create the color on the screen.

LIGHTNING AND FLAME

Plasmas are known to occur naturally on Earth under extreme conditions such as the heat created in lightning discharges and also in more mundane situations such as flames where some part of the material is ionized at all times. Still plasma is a relatively uncommon state of matter on Earth—at least compared with the rest of the universe, where it is considered to be, hands down, the most common phase for matter. For instance, stars consist mostly of plasma. Nebulae are interstellar clouds of gas and plasma.

Another naturally occurring plasma is the solar wind, a plasma spewing from the Sun in all directions. Evidence of the solar wind is found in the aurora borealis, or northern lights, the result of particles from the solar wind causing ionization in the atmosphere. The Earth's magnetic field naturally protects the planet from bombardment by the majority of these high-energy particles. As was shown by Hans Orsted (1777–1851) and Michael Faraday, electric fields and magnetic fields interact. A moving electric field will create a perpendicular magnetic field, and a moving magnetic field will cause an electric field to curve. The plasma released from the Sun follows the curve of the Earth's magnetic field, the magnetosphere, which effectively shields the Earth. Without this protection, the solar wind would cause immediate and complete erasure of all life on Earth.

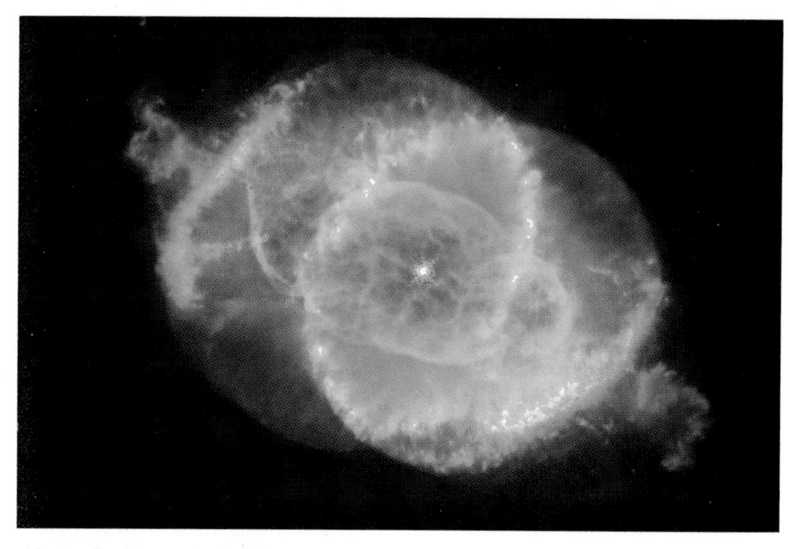

Above: Cat's-eye nebula. Plasma in the stars and in the space between them makes up the majority of the visible universe.

The phenomenon of aurora borealis, or northern lights, is caused by the solar wind, which is a stream of charged particles (plasma) that is ejected from our Sun and other stars.

PRESERVING OUR PLANET

There is no need to worry, however—unless one ventures beyond Earth's protective shield. There is much talk nowadays about the state of Earth's ecological health and how long humans will be able to survive as a species if a major climate change were to occur. Some have proposed that a solution might be to find and "terraform" another planet in another galaxy and colonize it. While the legion of practical problems associated with such an adventure is obvious, the present discussion demands the addition of one more. Not only must the candidate planet have a manageable temperature, soil to support plant life that could eventually create an oxygen atmosphere, and a gravitational field sufficient to hold this atmosphere, but the planet would also have to have a magnetosphere sufficient to ward off the solar storms that travelers would have to endure. A huge amount of study, innovation, and technology would be required to surmount this tremendous hurdle . . . so perhaps it would be best to use that same effort to save the Earth instead.

Water, a Necessary Compound

Why did *life*, the self-organizing, self-sustaining, self-perpetuating form of matter, spontaneously crop up on Earth? What were the conditions and the chemistry that gave rise to life? This question remains one of the fundamental mysteries of the universe. But much is known about the conditions necessary for life, and one of these is water.

WATER AND LIFE

Water, or H_2O, may seem a simple molecule, but a bit of reflection will show how very important it is to life and all human endeavors. A human will live only a number of days without water, and for other life forms the time may only be minutes or hours. What are the properties of water that make it so conducive to life?

The first can be understood in terms of the structure of water. Because of electron-pair repulsion, a water molecule has a V shape, with the oxygen at the apex and the hydrogens on either side, which lends it distinct characteristics that it would not have in any other shape. When water is heated up, the individual molecules begin to increase their motion and move in many different ways. They fly across the room faster, and because of its V shape, water can also vibrate—and rotate—and these additional modes of motion allow water to absorb a lot of heat energy without much change in temperature. Compare, for instance, the temperature increase of swimming pool water versus a cement sidewalk on a sunny day. The same amount of heat input from the sun raises the temperature of the sidewalk enough to encourage jumping in the pool.

How is this ability of water to resist temperature changes important to life? The human body, which is nearly two-thirds water, would not be able to survive one summer without the insulating abilities of water and would not survive the mildest winter, either. The chemistry of the human body is complex and delicate, as can be understood by considering the disastrous consequences of a disease-induced temperature rise of a few degrees. Yet humans are able to exist in climates as extreme as the Sahara Desert and the Arctic Circle—due to the body's wonderful insulating abilities.

When water is heated on a stove, hot water from the bottom of the pan rises, heating the water at the top, in a process called convection. Top left: Water is essential to all known forms of life.

The heat from the Sun will make this parking lot (left) unbearable on a summer day, while it raises the temperature of a swimming pool (right) a negligible amount.

LIQUID, SOLID, AND GAS PHASES

All phases of water are important to life cycles on Earth. When liquid water freezes, it expands with a force sufficient to create cracks in rocks. The eroding of rocks by water is an essential first step in liberating nutrients for the first link in the food chain, plants. The gas phase of water is critical to the movement of water from ocean to land in the water cycle. The liquid phase of water is able to dissolve solids, liquids, and gases, and the ability of watery human blood to dissolve nutrients enables cells to survive. Water's ability to dissolve oxygen allows fish to breathe, but not everything dissolves in water. Fats and oils separate from water, which again is an important consideration for life. The fats that form the walls of cells are congregated together by water, just as drops of oil will pool on a watery surface. Inside the oily bubbles that are cells, life processes are carried out in a watery, salty solution, much like the watery, salty solution of the ocean, the solution from which life likely began.

All water contains dissolved gases. The oxygen dissolved in seawater allows plants and animals to survive below the surface.

Solutions

The water matrices in which fish breathe and human blood cells swim are examples of solutions, mixtures that are uniform throughout. Solutions consist of a solvent, or carrier material present in the greatest quantity, and one or more solutes, the materials dissolved in the solvent. The air we breathe is a solution of oxygen, nitrogen, argon, and other gases. A solute in this case is oxygen, and the solvent would be nitrogen because it is the major component of air. Certain alloys can be described as solid solutions. Saltwater is an example of an aqueous solution, where water is the solvent. One of the most important parameters of any solution is concentration, the measure of the amount of solute in the solution. Concentration directly affects many of the uses and behaviors of solutions.

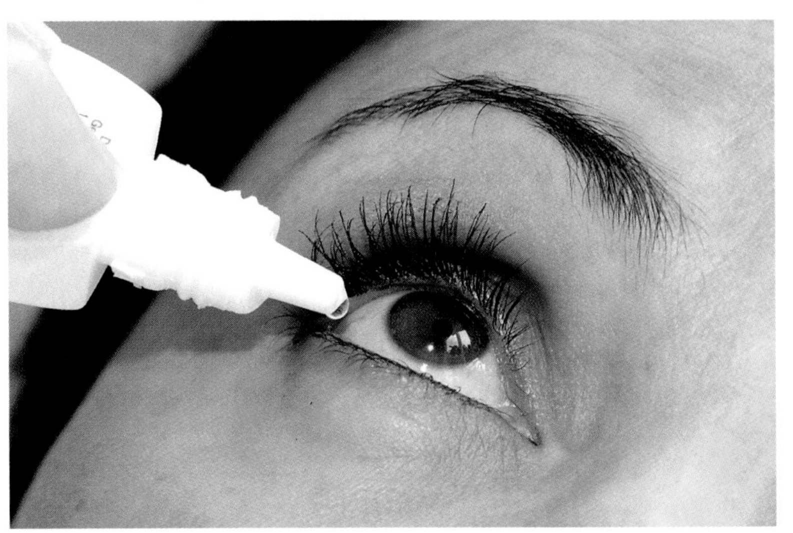

Eyedrops are made of a diluted saltwater solution that matches the concentration of natural tears. Top left: The process of osmosis regulates the movement of water into and out of living tissue. This process can cause skin to pucker after a bath.

COLLIGATIVE PROPERTIES

For instance, concentrated saltwater can be prescribed for a mouth rinse because a strong salt solution will kill many bacteria. Rinsing the delicate tissue of the eyes, however, requires the dilute saltwater found in commercial eyedrops. Commercial eyedrops are carefully formulated to be isotonic saline solutions, or solutions of the exact, physiologically correct concentration, because a different concentration could seriously disrupt bodily systems. The origin of such a disruption is part of an interesting category of solution behaviors called the colligative properties of solutions—that is, properties of solutions that are based on the collective effect of solvent and solute particles.

Colligative properties include freezing point depression, the tendency of a solution to freeze at a lower temperature than the pure solvent; boiling point elevation, the tendency of solutions to boil at a higher temperature than pure solvent; and osmosis, the tendency for solvent to flow into concentrated solutions—phenomena on which life processes are based.

FREEZING POINT DEPRESSION

Freezing point depression is responsible for the action of antifreeze. Water is used in radiators to conduct heat away from the engine, but if pure water were used, it might freeze during the winter and expand to break the radiator. When antifreeze is added, it lowers the freezing point of the water so that a much lower temperature is required to freeze the radiator.

Antifreeze, also referred to as engine coolant, protects cars from freezing in cold climates and from overheating in warmer areas.

Interestingly, antifreeze should also be referred to as antiboil because a solution will also boil at a higher temperature than the pure solvent, affording two automotive protections for the price of one!

SOLVENTS AND SOLUTES

The saying "learning by osmosis" describes the effortless flow of information into the memory; but while there is no such thing as effortless learning, there can be an effortless flow of solvent by osmosis. Solvent will flow from regions of low solute concentration to regions of high solute concentration through a semipermeable membrane, a membrane through which solvent can flow, but not solute.

For instance, osmosis is the process by which pickles pickle. Water flows out of the relatively dilute cells of the cucumber into the concentrated brine. Osmosis also causes skin to pucker after soaking in a hot bath. The cause is not the loss of water; it is the dilute water of the bathtub flowing into the more concentrated solution in the cells of the skin. The water-swollen skin folds in on itself. More critically, an imbalance of solute concentrations in bodily fluids can cause an osmotic pressure, which can contribute to high blood pressure, fluid retention, and other potentially harmful conditions. Osmosis is also the process by which bodily wastes are concentrated and toxins removed.

With the importance of solutions, the question arises: How do solutions form in the first place? For an answer, one can look once more at the electrons and how they spread themselves.

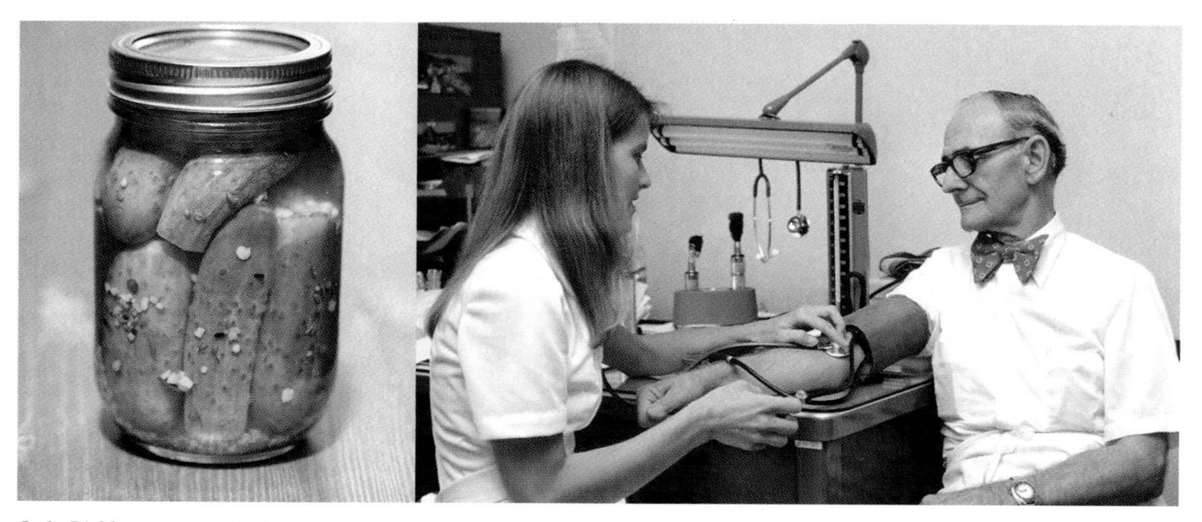

Left: Pickles are a result of osmosis. The concentrated brine replaces the water from the soaking cucumbers. Right: Osmotic pressure caused by an imbalance in bodily fluids can contribute to high blood pressure and other conditions.

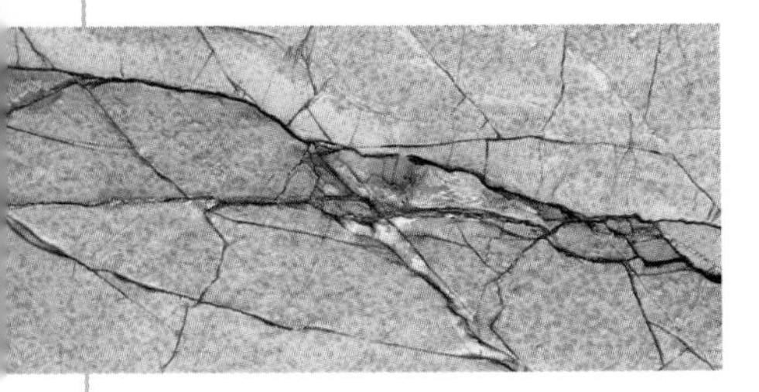

It's All in the Electrons

While it is convenient to depict electron orbitals as fixed regions of space—the billiard-ball model—orbitals actually more closely resemble a shape-shifting cloud. And just as a breeze can shape a cloud, so electrons in orbitals can shift in response to internal—and external—electric fields.

The V shape of water gives rise to a decidedly nonspherical arrangement of electrons, which creates an internal electrical field. Oxygen is more electronegative than hydrogen; that is, it has a greater attraction for the electrons of the molecule than does hydrogen. This unequal sharing of electrons results in water having one end that is more negative than the other, a condition called a dipole. Like two magnets, the more positive end of one dipole can be attracted to the more negative end of another, and this dipole-dipole attraction can be the force that allows water to condense. Forces felt between molecules—attraction or repulsion—are called intermolecular forces. Without intermolecular attractions, all molecules would be permanently consigned to the gas phase.

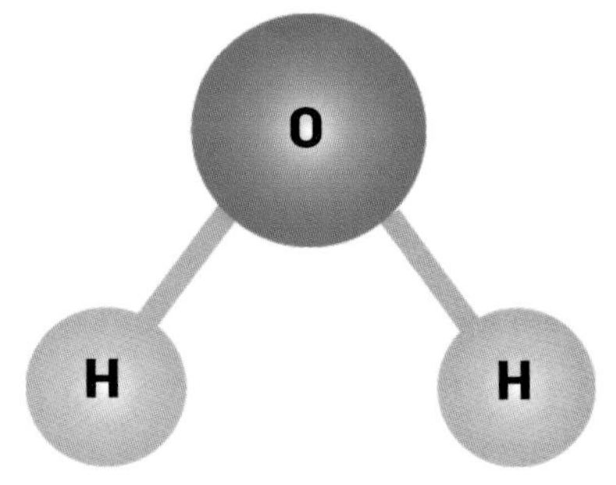

Above: A water molecule (H_2O). Top left: Ice can form in the cracks of rocks and expand, causing the rock to crumble.

Clouds are formed when water vapor condenses. The dipole-dipole attraction of water molecules allows condensation.

CONDENSATION

Water condenses quite readily because of its dipole, but molecules without a dipole can condense, too. Liquid nitrogen, condensed N_2, cannot have a permanent electrical dipole because there can be no uneven distribution of electrons between two identical nuclei. There can be, however, a momentary fluctuation in the density of electrons, and this fluctuation also can cause a fluctuation in a neighboring molecule. Because of this short-lived dipole-dipole attraction, called a dispersion force, even helium can be condensed if it is cooled to a low enough temperature.

HYDROGEN BONDING

Water, however, requires no extremes of temperature to condense, in part due to its internal

A liquid nitrogen (N_2) molecule has two identical nuclei and therefore can only condense as a result of a short-lived attraction called a dispersion force. This force is an induced dipole that is created.

dipole, but water also has the ability to engage in a special, strong intermolecular attraction known as hydrogen bonding. Hydrogen bonding in water is the attraction of the hydrogen end of one water molecule to the oxygen end of another. Other molecules that contain small, highly electronegative elements are also able to form hydrogen bonds, although the fact that water is able to hydrogen bond to itself through both arms of its V structure gives it several unique properties. For one, water expands when it freezes, whereas most other liquids shrink. This expansion is very important, because it allows ice to float or form in cracks of rocks, causing the rock to expand and be naturally crumbled into soil. Water expands because it needs to align each of its hydrogens with an oxygen, and vice versa, to optimize hydrogen bonding, just as a pile of building materials will take up more space when they are aligned in an open structure such as a house. Hydrogen bonds also contribute to the ability of water to dissolve many other compounds. Sodium chloride (table salt), for instance, is composed of positively charged sodium ions and negatively charged chloride ions. The positive end of water's dipole will be attracted to the negative ion, and the negative end of water's dipole will be attracted to the positive ion. As a result, water molecules insinuate themselves between the ions, forcing them to dissolve apart, and the hydrogen bonds between the water molecules build a solvent cage around the ions, keeping them apart.

DIPOLES AND DISPERSION

Dipoles, dispersion forces, and hydrogen bonding can bring electron clouds together, and if they approach each other with sufficient energy, there is the opportunity for the clouds to coalesce, to metamorphose into new patterns and form new molecules. This formation of new materials is the essence of chemical reactions, the subject that will be explored next.

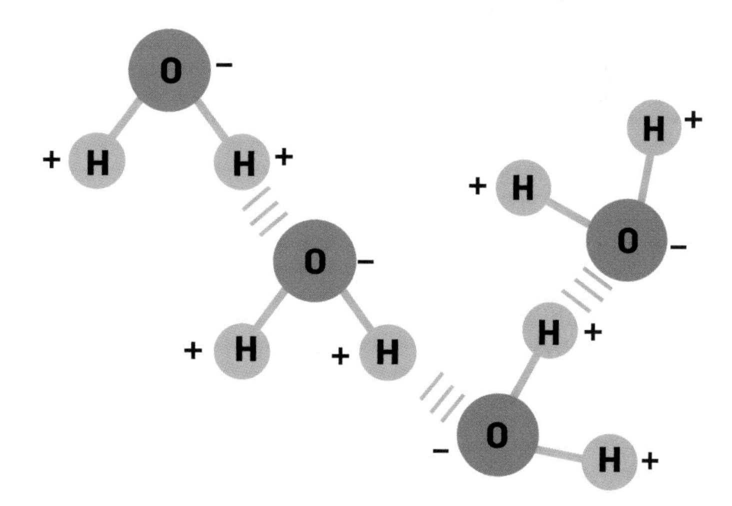

Hydrogen bonding is a force of attraction between a hydrogen atom in one molecule and an atom of high electronegativity in another molecule. This bond differs from other chemical bonds (ionic, covalent) in that it is intermolecular rather than intramolecular. Hydrogen bonding affects the properties of water and is important in proteins and nucleic acids. Hydrogen bonds also hold the two strands of the DNA double helix together.

CHAPTER 4

CHEMICAL CHANGES

Left: The Belousov-Zhabotinsky reaction under way in a shallow dish. This classic chemical reaction produces a dynamic display of concentric circles and patterns that oscillate over time as the compounds in the solution react. Top: Methane molecules from a gas line react with oxygen molecules in air to produce the heat and light of a stove-top flame. Bottom: The solid rocket boosters of a space shuttle are propelled upward by a powerful combustion reaction that generates aluminum oxide, visible as the familiar billows of white smoke that envelop the launch pad during the shuttle's takeoff.

We are surrounded by, and made up of, chemicals and chemical reactions. Such reactions generate electricity in the batteries that power a portable disc player. They allow us to breathe, to synthesize antibiotics, and to prepare appetizing foods. Some of these reactions can be observed, and marveled at, as they occur. For example, when a firecracker combusts, the reaction between oxidizing and reducing agents produces an explosion of brilliant light, color, and sound. Other chemically driven changes in matter, such as the formation of the enormous limestone tufa towers in Mono Lake, California, occur at a snail's pace over decades and even centuries. In this chapter, we will take a close look at how molecular reactions drive the rearrangement of matter, examine how combustion and decomposition attack molecular bonds, and discuss the properties of acids and bases. To better appreciate how the physical world is controlled by the behavior of molecules, we start by looking at what happens when atoms approach each other.

Chemical Reactions

Above: Fire is the result of oxidation—a chemical reaction that occurs when oxygen reacts with a combustible material, such as wood or gasoline, at a high temperature. Energy from the reaction is released as heat and light. Top left: The combustion of gasoline and oxygen in a car's engine generates energy that expands air within the cylinders and propels the car forward.

A simple but flawed analogy for a chemical bond might be the interaction between two magnets: if opposite magnetic poles are in the same vicinity, there will be an attraction that may pull them together to form a "magnetic bond." But chemical bonds are not just the result of attraction. When two atoms approach each other, the electrons of one may feel a pull from the positive protons of the other, but there will be repulsion, too. The negatively charged electron clouds will push away from each other, and the positively charged nuclei, when in close range, will feel a repulsion, too. If an arrangement of atoms can be found where these forces are balanced, then a chemical bond can form. However, unlike the magnets, the nuclei do not "touch" but are held apart by the complex tradeoff of forces that determine an optimum internuclear distance, termed the bond length.

But the truce is an uneasy one. The nuclei continue to vibrate back and forth around this equilibrium bond length, and with the right input of energy, vibrating nuclei can be made to vibrate apart, breaking the bond. If there are other nuclei nearby, and conditions are favorable, a new bond may form. The breaking of established bonds and formation of new bonds is a chemical reaction. Because of the fundamental complexity of this interaction of many protons and many electrons, chemists rely on experience as well as theory to predict the likelihood of a given reaction. With the categorization of reactions by common features, chemists have gained an understanding of the fundamental principles of chemical reactions and a powerful predictive tool.

PRODUCTS AND REACTANTS

To see how chemical reactions result in new materials, compare the properties of the reactants with the products. On the reactant side of the gasoline–oxygen reaction, for example, we have a highly combustible mixture of gasoline and oxygen, but on the product side we have carbon dioxide and water, two materials known for their abilities to extinguish fires. On the reactant side of the sodium–chlorine reaction we have a highly reactive metal, sodium, and a poisonous gas, chlorine, but the product, sodium chloride, is table salt, a stable consumable. Given the variety of chemical reactions—from life destroying to life sustaining—there is one principle that remains constant, and that is the conservation of mass.

REARRANGING MATTER

The conservation of mass maintains that matter is not created or destroyed over the course of a chemical reaction, but rather, rearranged. This principle is the basis for chemical equations—equations that show the rearrangement of reactants into products. To be valid, chemical equations must be balanced—that is—they must have the same amount of mass represented on either side. For instance, when one atom of solid sulfur reacts with a molecule of oxygen to form sulfur dioxide,

$$S + O_2 = SO_2$$

it can be seen from the balanced chemical equation that there is one sulfur nucleus in the reactants, and one in the products. There are two oxygen nuclei on the reactant side in the form of a diatomic oxygen molecule, and these show up as two oxygen nuclei in the product, SO_2 (sulfur dioxide). When the ratio of reactants is not one-to-one, coefficients are used in front of the formulas, as in the reaction of hydrogen and oxygen to form water:

$$2H_2 + O_2 = 2H_2O$$

From the balanced equation we see that two molecules of diatomic hydrogen react with one molecule of diatomic oxygen, for a total of four hydrogens and two oxygens on the left. These form two molecules of water, which are made up of two hydrogens and one oxygen each, for a total of four hydrogens and two oxygens on the right. Therefore, the mass of hydrogen is the same on the left and on the right, as is the mass of oxygen. The total mass is conserved. The equation is balanced.

A blue-green light is emitted when sulfur powder reacts with oxygen to produce sulfur dioxide, a malodorous, nonflammable gas that is a precursor to acid rain.

Both of the above examples would be categorized as synthesis reactions because they result in one product. Like other classification schemes, however—from biological to anthropological—it must be remembered that classification of chemical reactions is an attempt to impose order on nature, and nature often rebels. There is overlap between categories, and some reactions that are difficult to categorize completely. But a good deal of understanding can also be gained from categorization of balanced chemical reactions, as we will see in the pages that follow.

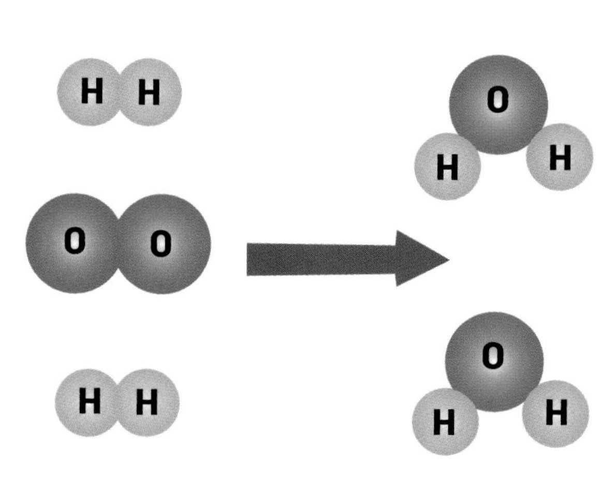

The figure above shows four hydrogen atoms reacting with two oxygen atoms to form two water molecules. Although the molecules are reconfigured, the number of atoms at the start and end of the reaction remains the same.

Chemical Synthesis

A chemical reaction is classified as a synthesis reaction when two or more reactants result in the formation of one product. The examples given in the last section—the synthesis of sulfur dioxide and the synthesis of water—fit the definition, as do subsequent reactions of the two products formed. Sulfur dioxide (SO_2) can react with oxygen gas in the air to form sulfur trioxide (SO_3), and then with water (H_2O) to form sulfuric acid (H_2SO_4). We will also see this kind of synthesis reaction again in chapter 10 when we discuss acid rain.

ORGANIC SYNTHESIS

Because of its role in bodily and medicinal chemistry, sulfur is also important in another class of synthesis reactions: organic syntheses. Organic syntheses, such as the synthesis of the sulfur-containing antimicrobial sulfa drugs, have changed the course of history. German biochemist Gerhard Johannes Paul Domagk (1895–1964) discovered that Prontosil, the first sulfa drug, was highly effective against streptococcal infections in mice. In an act of desperation, Domagk gave the drug to his daughter, who was suffering from a life-threatening streptococcal infection. His daughter recovered, and sulfa drugs went on to save countless lives in World War II. We will learn more about organic synthesis reactions in chapter 7.

PRECIPITATION REACTIONS

Chemical reactions that yield a solid product, or precipitate, in a solution are known as precipitation reactions. A common example of this kind of synthesis is the reaction between calcium ions and carbonate ions in water supplies. Hard water, which includes almost all natural water

Sulfur collection on glass jars in a volcanic field in Pozzuoli, Italy. Concentrated acids, such as sulfuric acid, are used in the manufacturing of fertilizers, dyes, and glue. Top left: In nature, calcium carbonate is found in eggshells, seashells, and limestone; many consumer products, such as chalk and antacids, also contain calcium carbonate.

For centuries, calcium-rich underwater springs and carbonate-rich water in Mono Lake, California, have reacted to create tufa towers, insoluble calcium carbonate (limestone) projections that can rise to over 30 feet.

sources, is water that contains impurities, most commonly an abundance of calcium and magnesium ions. Calcium ions will react with carbonate, which is also almost always present because of dissolved carbon dioxide from the air, to form scale. Scale, or insoluble calcium carbonate, leaves characteristic white deposits on plumbing fixtures, cooking utensils, and dishes. The intractability of scale can be discerned from another calcium carbonate compound, chalk. Anyone who has ever used chalk, and tried to wash it off his or her hands after use, can testify to its insolubility. Calcium carbonate deposits in water pipes and radiators can have devastating effects if the scale deposit is allowed to increase until it restricts water flow. Ironically, the same calcium and magnesium ions that can negatively affect our plumbing systems are in fact good for our health. Some of the undesirable effects of hard water, namely the scaly, white deposits on cutlery and fixtures, can be remedied by using water softeners that pump sodium ions that bond with calcium ions into the water supply.

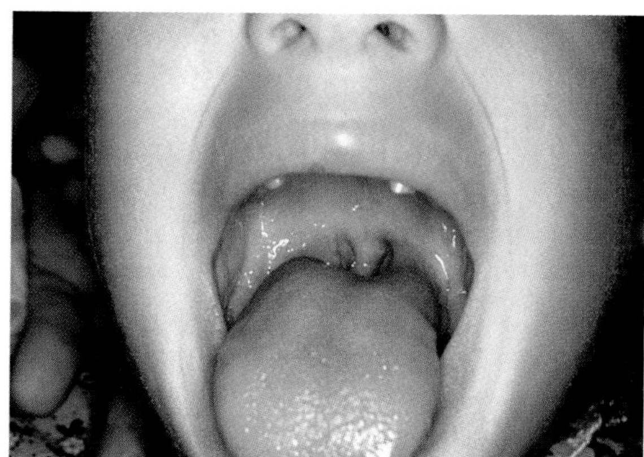

Sulfur-containing medications, first discovered by Gerhard Johannes Paul Domagk in the early 1930s, are still used today to treat bacterial infections such as streptococcus, fungal infections, and urinary tract infections.

Not all chemistry slows things down. For instance, oxidization and reduction mechanisms drive combustion and decomposition reactions, which attack molecular bonds and can produce dynamic, exothermic explosions.

Combustion and Decomposition

Combustion and decomposition reactions will be discussed together because they both break more chemical bonds than they make—and sometimes explosively. Most combustion and decomposition reactions are vigorous because the twin engines of energy and entropy drive them forward. The role of energy in bond formation is as we would intuitively imagine: Reactions are favored that release energy to the environment. Fires, explosions, and other energy-releasing reactions are commonplace and tend to be what we think of first when we imagine chemical reactions. It is hard to envision a Hollywood setting of a chemistry lab without boiling beakers and fuming flasks. But there are other perfectly legitimate chemical reactions that require an input of energy, such as cream of tartar with baking soda reaction. If cream of tartar and baking soda are dissolved in separate shallow glasses of water, allowed to settle and come to room temperature, and then carefully decanted into a plastic sandwich bag resting on

Above: Explosions occur quickly, releasing energy in a violent manner and generating heat and other combustible gases. Top left: A spectacular example of combustion at work, a basic firecracker is composed of oxidizing and reducing agents, coloring agents, binders, and regulators.

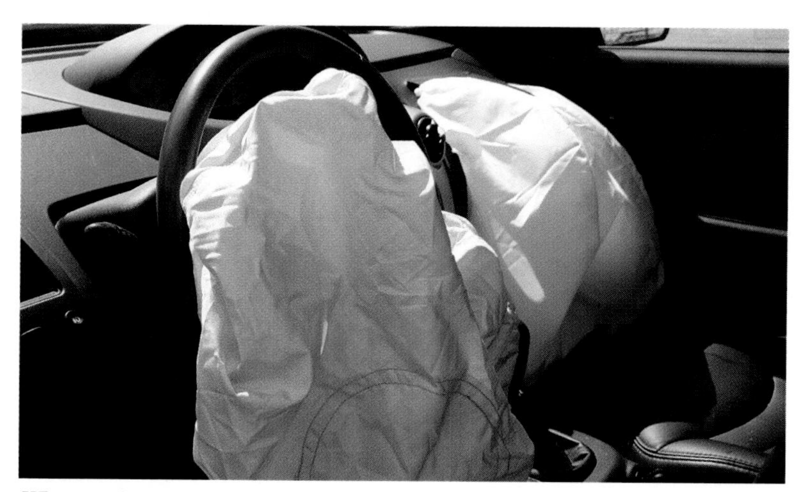

When crash sensors at the front of a car are activated, an inflator within the steering wheel produces an electric spark that ignites a stored reserve of sodium azide. The decomposition of sodium azide quickly releases nitrogen gas, inflating the airbag in 0.03 seconds.

the inside of a wrist, the wrist will detect a decided cooling—heat is being drawn in from the wrist to drive the reaction forward. So there must be another consideration besides the production of energy in successful reactions, and there is: entropy.

ENTROPY

Entropy is the tendency for natural systems to go toward a state of maximum disorder. For instance, a spoonful of salt tossed in the ocean can never be retrieved. The salt particles will spread and never come back together. A puff of perfume will dissipate; it will not locate in one nostril and remain. The explanation for entropy is statistical: there are so many ways of achieving a disordered state versus an ordered one that the disordered state is simply much more probable. For instance, the odds of dealing a perfectly ordered hand of bridge from a well shuffled deck, where every player gets thirteen cards of the same suit, is nearly 0.000 000 000 001, or essentially zero. Every other hand has a probability of nearly 0.999 999 999 999 . . . or essentially 1. A lot more probable. If you are dealing with any realistic sample of molecular-sized particles, then instead of 13 cards you have moles of molecules, and the odds for any ordered arrangement is much, much less. Consequently, the probability for a disordered collection is much, much better. Therefore, a reaction that increases disorder, such as an explosion or decomposition or combustion, is just more probable. Reactions that have both the energy and the entropy advantage have the greatest probably of occurring.

At this point we must mention that if a reaction is *predicted* to happen does not mean it *will* happen—at least on any observable time scale. But we will deal with the subtlety of reaction rates in chapter 5. For now we are talking about decomposition reactions that occur spontaneously—or need a little spark.

Decomposition reactions that require a little bit of a spark to get started include the reaction of gunpowder, the explosive reaction of hydrogen and oxygen, and one of the original reactions used to fill the air bags of cars: the reaction of sodium azide (NaN_3) to nitrogen gas (N_2). In a true testimony to the insistence of energy and entropy, once initiated, the decomposition of sodium azide can fill an air bag in a twentieth of a second, quickly enough to save many lives. And this life-saving reaction is necessary—because of the explosive reaction of gasoline and oxygen that provides the power for our internal combustion motors.

Hydrocarbons are compounds composed of carbon and hydrogen. Gasoline is made up mainly of hydrocarbons, and in their combustion hydrocarbons combine with oxygen to produce carbon dioxide, water, and heat. In the automotive engine, soot and other incomplete combustion products such as carbon monoxide are also formed, but if there is enough oxygen and controlled conditions, the combustion will be clean and the maximum amount of energy will be released. Oxygen promotes complete combustion because gasoline combustion is an *oxidation* reaction, one of the pair of coupled reactions that we explore next.

Oxidation and Reduction

Redox is a portmanteau word used by chemists. It combines the terms reduction and oxidation into "redox" because reduction and oxidation are coupled reactions—that is, there cannot be oxidation without reduction, and there cannot be reduction without oxidation. Oxidation reactions involve the loss of electrons, and reduction reactions require the gain of electrons. So oxidation without reduction would be an open circuit: electrons do not flow without someplace to go.

Consider the classic redox reaction between copper ion and iron:

$$Fe + Cu^{2+} = Cu + Fe^{2+}$$

This particular reaction was known to the ancients and formed a basis of the belief that base metals could be turned into gold. When iron is exposed to a copper ion-containing solution, the iron metal dissolves as the ion, as shown in the above chemical equation, and the copper ion deposits as a metal. The aspect that caught the alchemists' attention, and which they used as evidence for transmutation—the transformation of one element into another—was that the copper plated onto the remaining solid iron, giving the iron a shiny metallic yellow surface. If this reaction were observed as a natural occurrence for iron horseshoes retrieved from copper-rich streams, transmutation would seem a rational explanation! A more sophisticated understanding of the reaction, however, shows that it is a redox reaction because the copper ion has to gain two electrons, or be reduced, to become metallic copper:

$$Cu^{2+} + 2\ e^{-} = Cu$$

and electrons are provided by iron, which itself becomes the water-soluble ion:

$$Fe = Fe^{2+} + 2\ e^{-}$$

Iron is *oxidized* because it loses two electrons as shown in the above partial reaction, or half-reaction, as it is called.

A traditional mnemonic is "LEO the lion says GER": Lose Electrons in Oxidation; Gain Electrons in Reduction. But there is a historical origin for the terms, too. Consider the reaction between an iron oxide found in iron ore and carbon found in charcoal:

$$FeO + C = CO + Fe$$

This particular redox reaction has been carried out for many thousands of years in smelting furnaces from primitive to modern. The iron ore is crushed and heated with charcoal, which provides the necessary carbon, and the gas-phase byproduct,

Above: Modern chemistry builds on the protoscientific practices of alchemists, such as the one depicted in this nineteenth-century illustration. Top left: The oxidation of iron—when iron atoms react with oxygen atoms to form iron oxide—produces rust.

carbon monoxide, CO, removes itself by its volatility, leaving pure iron. The fact that an entire era is identified by this element, the Iron Age, attests to the importance of this technology.

The iron ore that is subjected to the smelting furnace contains other elements, minerals, and salts. Therefore it weighs more going into the furnace than when it is extracted as pure iron. The pure iron is *reduced* in mass; so the process became known as reduction. The paired reaction, oxidation, had to await its christening until the discovery of oxygen as a pure element in the 1700s, but even the ancients knew that charcoal had to be present for iron reduction. When it was discovered that the carbon from the charcoal gained oxygen in the iron-producing process, then the reaction of carbon became known as oxidation. We now know that electron-losing oxidations can even take place without oxygen and that the range of reactions, including organic reactions, that qualify as redox reactions is impressive. The same may certainly be said for our next class of reactions: acid–base reactions.

When fossil fuels, such as coal (shown left), are burned, they react with oxygen and release carbon dioxide, water, and energy. About 85 percent of the energy we use comes from fossil fuels.

Acids and Bases

Most people have an intuitive notion of what constitutes an acid, if not from direct experience with an acidic material then from the Hollywood horror cliché of a bubbling, corrosive acidic brew. Indeed, most acids, under the right circumstances, can be corrosive. On the other hand, our stomachs contain an acid solution, cream of tartar is an acidic salt, and coffee is considered slightly acidic, too, so bubbling and boiling are obviously not necessary. The best way to tell if a solution is acidic is to test it with another substance called an indicator.

An indicator is a chemical that indicates the presence of acid, usually by a color change. As early as the 1600s European chemists knew that extracts of highly colored flowers and vegetables could act as indicators. Purple cabbage, finely chopped and allowed to soak in water, with perhaps gentle warming, makes a purple solution that is an excellent indicator, as does a bit of red wine in water. Vinegar, an acid solution, will change the color of the purple cabbage solution bright pink. Vinegar will also change the color of wine, but the color change will depend on the wine. Litmus paper is paper that has been treated with an indicator, and one good definition of

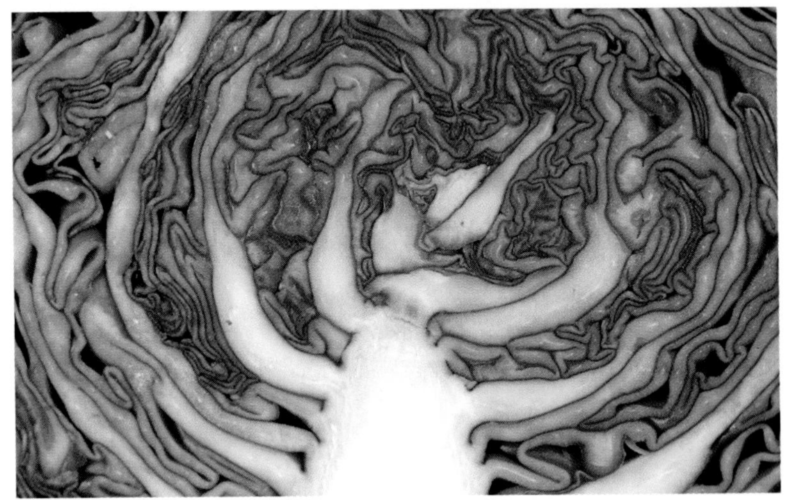

Above: Make a simple pH indicator by boiling half a head of finely grated red cabbage. The anthocyanins in the cabbage juice react with acidic and basic solutions and turn colors—neutral solutions will turn purple, basic solutions will become green-yellow, acidic solutions will result in a red color. Top: When you eat a lemon or lime, citric acid in the fruit reacts with protein molecules on your tongue. This reaction sends an impulse to the brain, which allows you to interpret the taste as sour or tart. Top left: Pickling cucumbers, peppers, cabbage, and other low-acid foods in jars with acidic solutions helps them resist spoilage because microbes cannot survive in the acidic environments.

Dipping a strip of litmus paper into a solution containing a common laboratory acid, such as sulfuric acid or hydrochloric acid, will cause the strip to turn red. Conversely, when dipped into a basic solution, the litmus paper will turn blue.

an acid is a solution that turns blue litmus paper red. But for coffee, which turns everything brown, this definition might fail, so we need a more universal definition. The one that is most often used is that an acid is a substance that reacts with a base. So what is a base? A base, of course, is something that reacts with an acid.

Most find the above rather unsatisfying because of the circularity and because these definitions beg a more familiar definition of a base. We can provide some general observations concerning bases—such as bases tend to taste bitter and feel slippery and turn red litmus blue—and we can provide some common examples of a base, such as household ammonia and baking soda. We can even test for bases using our purple cabbage or wine indicators. A solution of baking soda in water will turn our purple-cabbage indicator teal blue. Household ammonia in our purple-cabbage indicator will turn the solution green, and ammonia in our red-wine indicator will turn the solution dark blue-green.

But ultimately no characterizations are as useful as the definition that an acid is a material that reacts with a base, and a base is a material that reacts with an acid. Perhaps the reasons might be better understood by considering what we call this acid–base reaction: It is called neutralization.

Neutralization is again a term for which we have a Hollywood metaphor: To neutralize a rival, in popular parlance, means to render someone ineffectual. And this is, fundamentally, the interaction between an acid and a base: an acid will neutralize a base and a base will neutralize an acid. To understand the chemistry of the neutralization reaction, we will begin with the simplest definition of an acid and a base. In this definition, called the Arrhenius definition, for Svante Arrhenius (a chemist we met in chapter 2), we identify an acid as a chemical species that donates H^+ to the solution. A base, according to the Arrhenius definition, is a substance that donates OH^- to the solution. The neutralization reaction, then, would be the marriage of acid and base, H^+ and OH^-, to form water:

$$H^+ + OH^- = H_2O$$

While there are common hydroxide-containing bases, such as sodium hydroxide, or lye, which was used for many years as a crystalline drain cleaner, the Arrhenius definition is too narrow to include examples of base that we used with our indicator: ammonia and sodium bicarbonate. Therefore chemists, most notably Johannes Nicolaus Brønsted (1879–1947) and Thomas Martin Lowry (1874–1936), recognized the need for a broader definition. The Brønsted-Lowry definition identifies a base as a material that, like OH^-, accepts H^+, such as ammonia and the bicarbonate ion. At about the same time, G. N. Lewis (whom we also mentioned in chapter 2), proposed a definition of acids and bases that was broader still, but the definitions we have developed so far are sufficient to lead us to a discussion of pH, an important concept in acid–base chemistry, on the following page. We will examine the pH scale, which is used to measure the acidity or alkalinity of substances, in the context of chemical reactions, and observe how pH is important to such ordinary concerns as skin care, swimming pools, and aquariums. In fact, there are instances of all the classes of chemical reactions that are part and parcel of our everyday lives.

Constantly Changing

We cannot escape chemical reactions in our daily life—nor would we want to! We are kept alive by a myriad of ongoing bodily chemical reactions, and we are kept warm, well-fed, and comfortable by as many more. In fact, a system that provides examples of all the classes of chemical reactions we have discussed so far is something found right in many back yards: a swimming pool.

For anyone who has ever maintained one, the mention of chemicals in connection with a swimming pool is certainly no shock: Swimming pool chemicals are an absolute necessity for proper care and maintenance. Parameters such as total alkalinity, chlorine content, and pH have to be monitored on a regular basis.

The term "pH" is as familiar as it is mysterious. The pH of an aquarium must be kept at the correct value to support aquatic life. The pH of soil must be carefully controlled if pink hydrangeas are desired instead of blue. Personal-care products may be promoted as

Above: Because sudden or drastic changes in pH are common causes of fish fatalities, regular monitoring of these levels is necessary. Top left: Hydrangeas grown in aluminum-rich soil react to the acidic environment by taking on a blue or purple color. Depriving the hydrangeas of aluminum by planting them in less acidic (or more alkaline) soil will produce pink blooms.

"pH balanced"—but the careful consumer might wonder: What exactly is pH?

The pH of a solution is a measure of its acid level, but on a peculiar kind of a scale: The higher the concentration of acid in solution, the smaller the number on the pH scale. For instance, lemon juice, which we all agree is fairly acidic, is about 2.3 on the pH scale. The pH of a soap, which gives away its identity as a base by its bitter taste, might be on the order of 10 or so. Acid content is measured on a pH scale for the same reason sound is measured on a decibel scale. Every time the sound level or acid level makes a 10-fold change, the decibel or pH scale changes only by 1. Using this type of scale, a *logarithmic scale*, allows values that differ by many multiples of 10—as from millionths to tenths—to be placed on the same scale.

The Maillard reaction (also known as the browning reaction) occurs when meat is cooked at 300°F to 500°F, causing sugar and protein molecules on the surface of meat to react and produce a "meaty" flavor and brown color.

A pH of 7 is neutral, and this would be the pH of pure water. But the skin has a slightly acidic pH, which is why personal care products are pH balanced. One of the decomposition products of the waste produced by animal life in all forms is the base ammonia, so pH levels must also be monitored for a healthy aquarium. In swimming pools, pH levels are very important: A swimming-pool solution that is too basic will promote the formation of deposits, precipitates, that could clog the water-circulation system. Sodium hypochlorite, which provides the active oxidizing agent in household liquid bleach, is the antibacterial, anti-algae, antifungal agent of choice for swimming pools. But hypochlorite levels must be carefully monitored, too, because hypochlorite continuously undergoes decomposition.

Swimming pools, of course, are not the only places where common chemical reactions occur in our lives. Cleaning involves chemical reactions, as does automotive maintenance, lawn care, and plumbing. Chemical reactions are also involved in cooking and in metabolic processes. In these last two examples, however, we introduce a new component in the chemical mix: energy. Cooking requires the input of energy for the desired result, and the energy output of metabolism keeps us warm and alive. The central role that energy plays in chemical reactions is the topic we explore next.

CHAPTER 5

ENERGY

Left: Waterfalls, such as the Fairy Falls in Multnomah County, Oregon, are one of nature's pristine examples of the conservation of energy. Water at the top of the waterfall has potential energy. As it falls, the water is converted to kinetic energy, and when it hits the ground it is converted to thermal energy.
Top: At present, typical solar cells are not capable of efficiently harnessing energy from the Sun, although more-practical technologies related to solar energy are being studied.
Bottom: Wind is a clean, efficient, and renewable source of energy, but because it is unpredictable, an additional energy source or energy storage system needs to be used as an adjunct.

Energy is all around us, even in the least technological of settings. The *potential* energy of water pooling at the top of a waterfall is transferred into *kinetic* energy as it begins cascading downward. Energy's many forms allow us to do more work on a larger scale than the generations that have come before us. Yet with all the tremendous achievements and advancements that technology has provided, the release and transfer of energy—be it potential or kinetic, heat or radiation—boils down to the interaction of particles we will never see. The energy it takes to fire a rocket into orbit begins at the subatomic level. The energy that lights up a baseball stadium is first generated by impossibly tiny electrons migrating from atom to atom. And simply splitting one tiny atom in two can result in the release of a catastrophic amount of energy capable of destroying an entire city or more. Thankfully, however, the most fascinating developments in energy are leading us back to an appreciation of the natural world from which energy originally springs. Sun, sky, wind, and water will someday power our grateful planet.

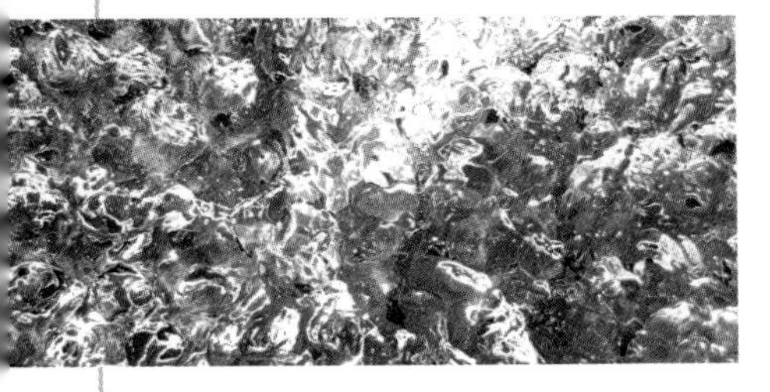

Heat and Energy Transfer

Energy, often described as the ability to do work, exists in many forms. We see energy in action all around us: cars driving, people jogging, the Sun shining. Heat, or more precisely, thermal energy, is one form of energy that is rather familiar to most people. Everyone, for example, has seen heat applied to a pot of water on a stove. As the temperature of the water increases, the heat that has been applied causes the water to boil. The steam rising from the boiling liquid is a visual affirmation of the transformation of liquid water into a gaseous form, or steam.

But water boiling on a stove is just a tangible example of what heating the water causes. The real action involved takes place far beyond the scope of human observation, at the molecular level. Heat is the internal energy of a system—the sum of the potential and kinetic energies. Temperature is one way to measure the amount of thermal energy that a particular material possesses. The higher the temperature of a substance, the more excited and in motion its particles are. When water boils, the thermal energy of the system is raised by the addition of a heat source. When the temperature, the measure of heat energy, reaches 212°F (100°C), a phase change begins to take place. The kinetic energy of the water molecules becomes great enough to allow the molecules to break free of their bonds to one another and, eventually, the water's surface.

Above: The transfer of heat by convection drives circulation currents in hot springs, such as the Morning Glory Pool in Yellowstone National Park. The temperature gradient within the spring makes it possible for thermophilic bacteria to thrive in the hot center regions, giving the water a distinct blue-green color; different-colored bacteria populating the cooler, outer edges of the spring create the yellow-white tint. Top left: When water is heated to its boiling point, 212°F (100°C), steam rising from the boiling indicates that a phase change is occurring.

GETTING FROM HERE TO THERE

The law of the conservation of energy states that energy is never lost. In an isolated system, the total amount of energy remains the same. It's not created. It's not destroyed. Though the amount of energy remains constant, it can be changed from one form of energy into another. And energy, including heat, can also be transferred. The transfer of thermal energy goes from areas of higher temperature to lower temperature. This transfer can occur in three basic ways: convection, conduction, and radiation.

CONVECTION & CONDUCTION

The transfer of heat by the flow of liquids and gases is called convection. Returning to the water on the stove, as the water closest

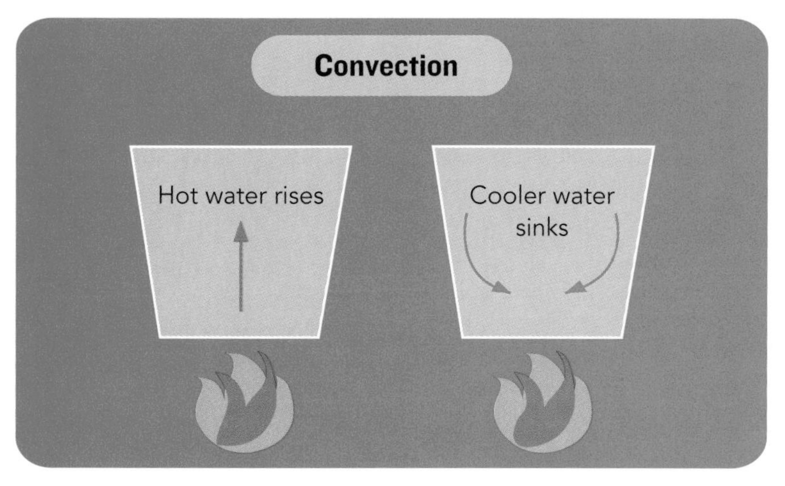

As water is heated in a pot on the stove, the liquid closest to the heat source becomes warmer and, because of the increase in molecular motion, expands and becomes less dense. The heated water then rises toward the top of the pot, circulating the denser, cooler water downward, toward the heat source, where it in turn is heated.

This false-color image of the Earth shows relative amounts of heat emanating from the top of the atmosphere in the form of long-wave radiation. The thermal energy leaving the oceans is mostly uniform, whereas in the American Southwest (upper right-hand corner), there is an increased amount of outgoing radiation as a result of the limited presence of water to absorb solar energy.

to the heat source becomes warmer it also becomes less dense and rises toward the top of the pot. The denser, cooler water moves toward the bottom. As the warmer liquid encounters the cooler liquid, heat is transferred upward through the water—always going from areas of higher temperature to areas of lower temperature. The same convection phenomenon happens with heated air in a room and also, on a much larger scale, within Earth's atmosphere.

Conduction, on the other hand, occurs when heat is transferred from atom to atom, molecule to molecule. The atoms and molecules collide with each other, and the transfer of kinetic energy—the energy of movement of the atoms or molecules—takes place. Different substances are capable of different levels of heat conductivity. For example, if the pot on the stove is made of aluminum, it will conduct heat much better than if it were made of iron. The heat received at the bottom of the pot closest to the heat source is conducted throughout the entire body of the pot.

RADIATION

The clearest example of the transfer of thermal energy by means of radiation is sunlight. Radiant energy from the Sun travels over 91 million miles to reach Earth. Solar energy, a type of radiant energy, travels in electromagnetic waves and does not need a liquid, gas, or solid to carry it, as in the cases of convection and conduction.

Getting Things Started: Activation Energy

Dynamite explodes. The copper surface of the Statue of Liberty slowly turns green after years of exposure to the elements. These phenomena are examples of chemical reactions taking place. Some of these reactions happen at lightning speed, while others need to take their time, be it hours, days, or even years. Many factors can affect the rate at which a chemical reaction occurs. And many chemical reactions, no matter how quickly or slowly they take place, need a little boost to get started. That boost is called activation energy.

Copper oxidizes when exposed to oxygen. Over a period of decades, the Statue of Liberty has slowly changed from penny-colored to light green.

Imagine a car with the motor running. In order to get the car's motor started, the ignition must be activated. Activation energy is the minimum amount of energy needed in order to start a specific chemical reaction. Once reactions have gotten over that hump and have achieved their activation energy, reaction rates vary greatly. These rates can be affected by things like temperature and pressure, or by the concentration of the liquids or gases that need to react. Controlling these rates is an important part of chemical manufacturing.

SPEEDING THINGS UP

Like the single spark that starts the inferno, very small things can cause a drastic change in the rate of a reaction. And in many industrial applications where chemical reactions are used, it is useful to be able to help reactions along and speed them up a bit.

The energy we call heat can be applied to reactants to help speed up the rate at which they react with each other. In order for reactants to react with each other, their molecules need to come in contact and they have to collide with enough energy. As the molecules of the reactants heat up, their energy of movement, or kinetic energy, increases. This added

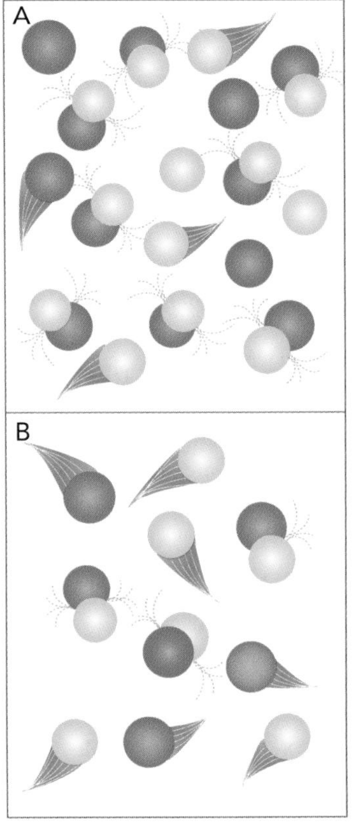

Above: Atoms and molecules, which are always in motion, move even faster when heated. For example, the molecules in boiling water move relatively quickly (Figure A) compared to water molecules at room temperature (B).
Top left: The detonation of dynamite releases a tremendous amount of energy, making it an exothermic reaction.

speed can increase the frequency with which the reacting molecules will collide and speed up the rate of their reaction. Changing the concentration of reactants, in the case of liquids, for example, can also increase the rate at which reacting molecules will collide with each other and increase the overall rate of reaction.

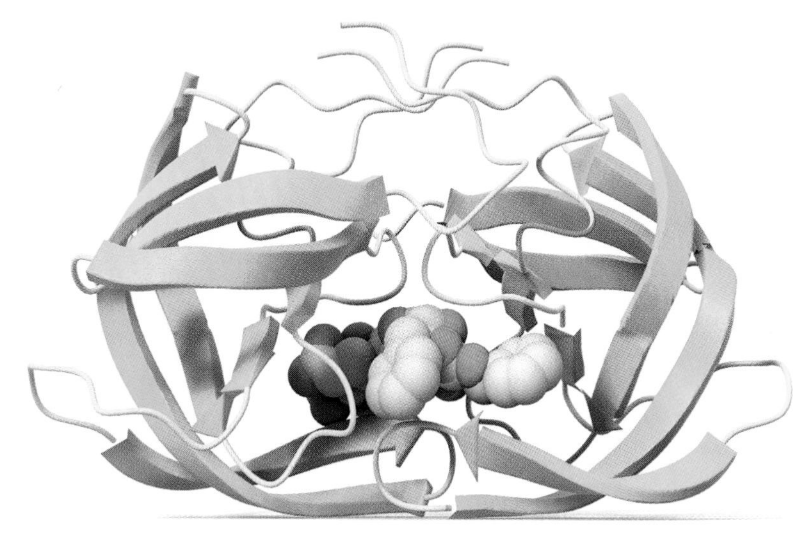

In 1989, scientists discovered the three-dimensional structure of HIV's protease—the enzyme the virus needs in order to make more copies of itself. This computer-generated image shows HIV's protease inhibitor, an antiviral drug that disables the viral enzymes by creating a block in the activity site at the center of the protease.

A catalyst can also be used to speed up a reaction without changing the reactants, or the product of the reaction. A catalyst can bind with one of the reactants, making it simpler for another reactant to react with it. It gives reactions more options, in a way. In bonding with reactants, the catalyst provides them a different, shorter route to get to the same destination or product. Aluminum chloride is a very common and useful catalyst involved in a wide variety of processes in the chemical manufacturing industry, from rubber to pharmaceuticals. Aluminum chloride works by helping to combine and rearrange the lengthy carbon chains in organic molecules so that they can then form new organic compounds. Shifting the focus a bit closer to home, enzymes are biological catalysts that help maintain the innumerable chemical reactions that take place throughout our bodies every second of every day.

SLOWING THINGS DOWN

Everyone is in a big hurry today. But sometimes it helps to slow down. In chemical reactions, it is often more advantageous to be the tortoise rather than the hare in the race to form a new compound. The ability to control the rate of a reaction can be as important as the results. Inhibitors, the functional opposite of catalysts, do just what their name indicates: They inhibit reactions, causing them to take place at a slower pace and sometimes stopping them completely. One way inhibitors work is to cut down on enzyme activity, preventing enzymes from binding to their specific substrate. This is useful in pharmaceutical applications. In fact, some of the drugs used to treat HIV are viral enzyme inhibitors.

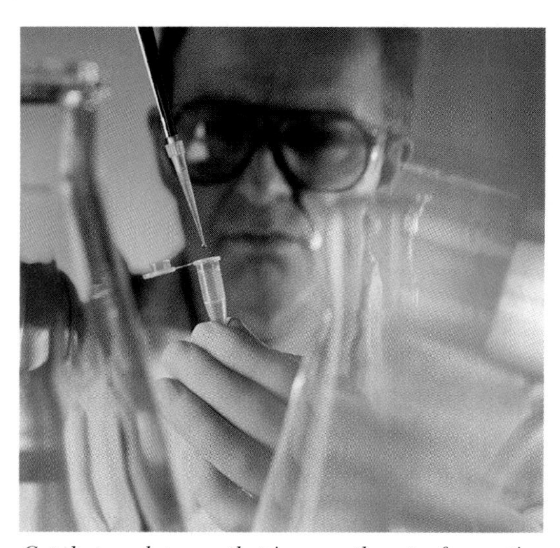

Catalysts, substances that increase the rate of a reaction by lowering its activation energy, are used extensively in research for many industrial applications.

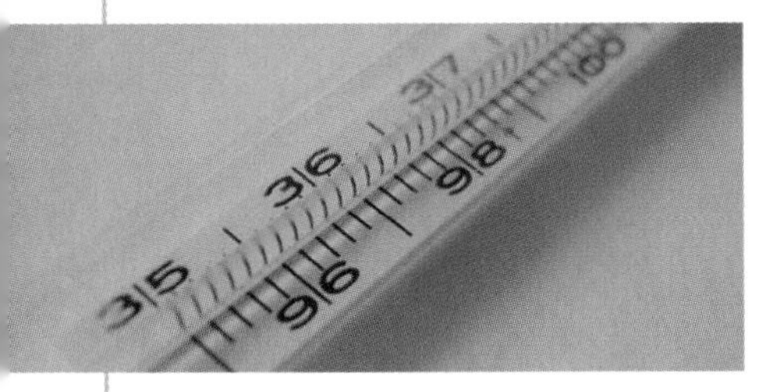

Temperature Scales

Often used interchangeably with the concepts of heat and cold, temperature is much more. Heat and cold are relative, and while talking about the weather may involve discussions of the temperature, what temperature really measures is the average thermal energy, the kinetic energy of the molecules of a particular substance.

Temperature makes no sense without the proper scale. Saying, for example, that someone or something weighs "145" is meaningless without indicating the scale: pounds, kilograms, ounces, or the like. A number of temperature scales have been invented over the last few centuries and are still in use. The importance of these scales was that they went beyond the ability to show that temperature differences exist—as Galileo's first thermoscope did—to actually measure those differences. The point at which water boils and freezes are the two key reference points in the major temperature scales.

FAHRENHEIT SCALE

The Fahrenheit scale is named after Gabriel Fahrenheit (1686–1736), who set the freezing point of water at 32 and the boiling point at 212. Fahrenheit is perhaps the most familiar

Above: Liquid thermometer. Volumetric changes of the liquid in the thermometer indicate temperature variations. Left: Scientists often rely on the Celsius scale, which assigns 0°C to the temperature at which water freezes, and 100°C to the temperature at which it boils. Top left: The reading on an average thermometer indicates the average kinetic energy, or energy of motion, of the particles being measured.

temperature scale to those living in the United States. It is the scale used to refer to outside air temperature and the scale used to measure the temperature inside ovens. Fahrenheit is also credited with creating the first mercury thermometer.

CELSIUS SCALE

Not long after Fahrenheit came Anders Celsius (1701–44) who created the scale named after him. Celsius set his boiling point of water at zero degrees and his freezing point at 100, thus creating increments of temperature change different from those of the Fahrenheit scale, which has a 180-degree difference between its freezing and boiling points. After Celsius died, so did an aspect of his scale. The reference points were reversed—zero became freezing and 100 became boiling—and the name of the scale was changed to centigrade in reference to the 100 increments separating the freezing and boiling points. In the mid twentieth century, Celsius again became the name of choice, and his is the scale commonly used in chemistry, physics, and all scientific disciplines. It is also used in day-to-day life around the world to measure air temperature.

KELVIN SCALE

The Celsius scale is also defined as the Kelvin scale minus 273.15. The Kelvin scale was named after William Thomson, Lord Kelvin, who devised this absolute temperature scale when he was not yet 25 years old. The scale uses absolute zero as its chief reference point. Absolute zero is exactly what it sounds like—it's as low as it gets. There is no colder temperature. Defined as zero degrees Kelvin, absolute zero is a theoretical temperature at which no measurable heat or thermal energy would be present and atomic thermal motion would grind to a halt. Molecules and atoms always possess some measure of thermal energy, even at extremely cold temperatures. A substance at absolute zero would have the lowest possible amount of molecular energy, and that energy could not be transferred to other systems. Absolute zero has never been achieved, although scientists have come very close—within billionths of a degree—under laboratory settings. The coldest temperature ever recorded on Earth was -128.6°F (-89.2°C) at Vostok, Antarctica.

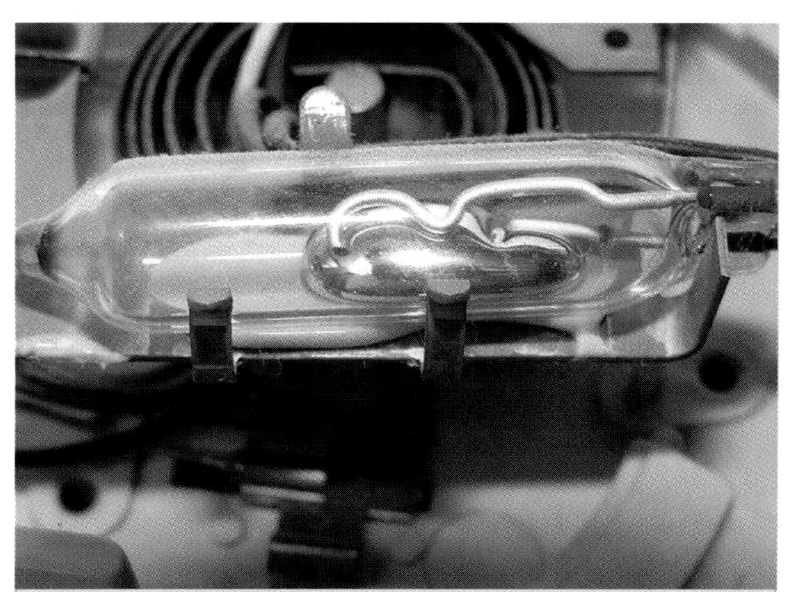

Exposure to inorganic contaminants, such as the mercury found in an old air conditioning thermostat, can damage the kidneys and central nervous system.

AN ODE TO QUICKSILVER

Mercury is an element with a wide variety of unique properties. It is the heaviest known elemental liquid, and not only is it the only metal that is liquid at room temperature, but mercury remains in a liquid state over a wide temperature range, from its melting point at roughly -38.2°F (-39°C) to its boiling point at 674.6°F (357°C). Combine this with mercury's overall sensitivity to temperature change, and this slippery, silvery element makes a very useful ingredient in thermostats and thermometers. At least it *did*.

Mercury's rather unattractive qualities—primarily its toxicity to living things—have resulted in a phasing out, even an outright ban, of mercury in general medical devices and home-use thermometers. Modern, over-the-counter thermometers now do the same job with alcohol and red dye.

Energy from the Sun

Earth's biggest power plant is not on Earth at all. It is at the center of the solar system. Our closest star, the Sun, functions like a giant nuclear reactor. On Earth, we burn gasoline and other fuels to run our cars and homes. The Sun runs on hydrogen, the most abundant element in the universe. The Sun creates its energy by performing the same chemistry experiment over and over again: It slams four hydrogen nuclei together to create a single helium nucleus. The excess energy produced by this nuclear reaction is so great that it shoots out of the Sun and radiates outward into the universe, warming the Earth and other planets.

Fusing together the nuclei of two or more elements is no small feat, but scientists have been able to make fusion occur, from early accelerator experiments in the 1930s to hydrogen bombs and the first attempts at fusion reactors in the 1950s. However, scientists have not been able to make a controlled reactor that generates more power than it takes in. Because fusion creates more energy from relatively little matter, the Sun is, in a sense, a far more efficient power plant than any of the ones humans have created. Of course, the Sun has a huge advantage over man-made power plants or even fusion laboratories: It is extremely, large, hot, and dense. The Sun's surface alone is about 10,000°F (6,000°C). As hot as that sounds, it's nothing compared to heat at the center of the Sun. The Sun's core, the power plant itself, may be as hot as 27,000,000°F (15,000,000°C).

Think of the work that much heat can do! In one second, the Sun can fuse 600 million tons of hydrogen into 596 million tons of helium. The four-million-ton leftover is converted into the energy that shoots into space. When that happens, in a second, one square inch of the Sun's surface can emit an amount of light that is equivalent to 400 100-watt lightbulbs. Of course, not all that energy reaches the Earth. Scientists estimate that about 1 part in 2.3 billion of the Sun's light reaches Earth and 1 part in 3 billion makes it all the way to the Earth's surface. The rest of its light is not pointed toward us, and dissipates into space.

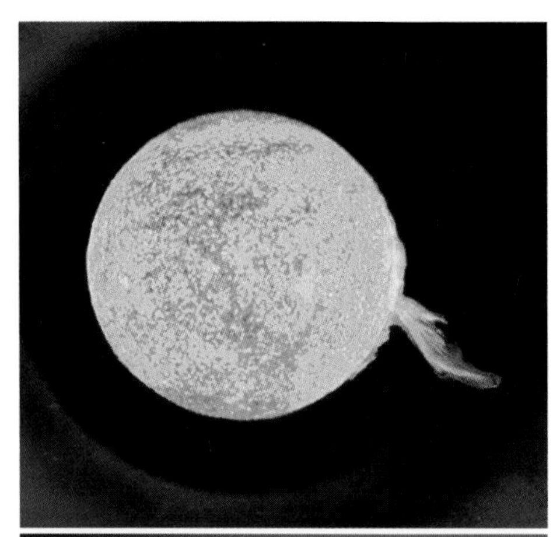

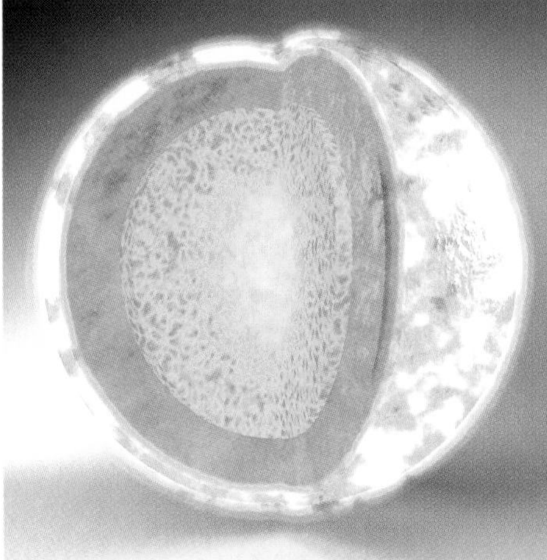

Above: The Sun's core (bright white) is the hottest and densest region, where nuclear reactions that convert hydrogen into helium produce light and heat. Energy produced in the core travels through the radiation zone (yellow) until it reaches the convection zone (orange), where thermal convection transfers energy to the Sun's surface. Top: The Sun's surface is called the photosphere, because it emits the visible light that reaches the Earth. Top left: The Sun has been burning for billions of years and is Earth's fundamental energy source.

During photosynthesis plants absorb sunlight, water, and carbon dioxide and expel oxygen while they produce sugar, which they need to survive. Without the Sun's light, there would be no oxygen.

POWERED BY THE SUN

Life on Earth would be impossible without this energy from the Sun. Sunlight and warmth are the impetus for nearly all the Earth's important processes. Photosynthesis is just one example. Plants convert the Sun's photons, or particles of light, into chemical energy. Leaves absorb sunlight, water, and carbon dioxide and produce the sugar glucose, which the plant "eats" to survive. The plant expels oxygen as a by-product and evaporates water (transpiration).

The Sun also drives the Earth's hydrological, or water, cycle. Heat from the Sun warms the oceans, which causes water to evaporate off the ocean's surface as gas or steam. That gas rises upward, forming clouds, and as the water vapor trapped in those clouds cools down, the gas condenses into water and falls back to Earth as precipitation, either as rain, snow, sleet, or hail. Precipitation may be absorbed by the Earth and flow down into underground water reserves. Or water may simply flow across the surface of the Earth in rivers and streams until it reaches larger bodies of water or the ocean itself, where the whole process begins again. All the while, humans and animals are exhaling water vapor in a process called respiration, and plants continue their process of transpiration. This vast cycle provides the Earth's inhabitants with the nutrients and liquid they need to survive. It would come to a painful stop if the Sun stopped shining tomorrow.

Many people would like to see humans more directly harvest the energy of the Sun to power homes, businesses, and cities. By one estimate, in a single day the Earth receives more energy from the Sun than humans could consume in 27 years—and yet, much of that raw energy is not so much wasted as ignored by the Earth's people. Later on in this chapter, we'll look at how solar power can be harvested as an alternative energy source.

The Sun also drives the Earth's water cycle—warming the oceans and causing water to evaporate off the ocean's surface as gas or steam—ultimately providing plants and animals with nutrients and liquids essential for survival.

Fossil Fuels

Fossil fuels such as natural gas, coal, and petroleum are the remains of plants, animals, and microorganisms that flowered or scurried across the Earth in the late Paleozoic. When the plants and animals died more than 250 million years ago, their remains fell to the bottoms of lakes, swamps, and oceans, and broke down into a spongy material called peat. Over time, the peat was covered by still more layers of sand and rock. As the weight of these extra layers built up, water was squeezed out of the peat so that only the material's carbon essence remained.

The results of this compression lay undisturbed in the Earth for eons, until the birth of the Industrial Revolution created a need for copious amounts of fuel to run factories. Humans began to drill through layers of rock to extract oil. They chipped their way into mountain caverns to find rich layers of coal. And they punched through promising layers of stone to capture natural gas. These fuels are also called by their chemical name, hydrocarbons, because they consist of the elements carbon and hydrogen.

These compounds contain a lot of trapped energy, which can be released only during a reaction that is strong enough to break their chemical bonds. The chief method of breaking those bonds and releasing that energy is combustion, or burning. In an ideal combustion situation, one uses a small amount of energy to bring about the release of a much larger amount of energy. Think of how a car's internal combustion engine works: A tiny spark from the spark plug ignites a volatile mix of gasoline vapor and oxygen; the result is a much stronger explosion, water vapor, and exhaust.

In theory, fossil fuels burn relatively cleanly because common hydrocarbons, such as octane,

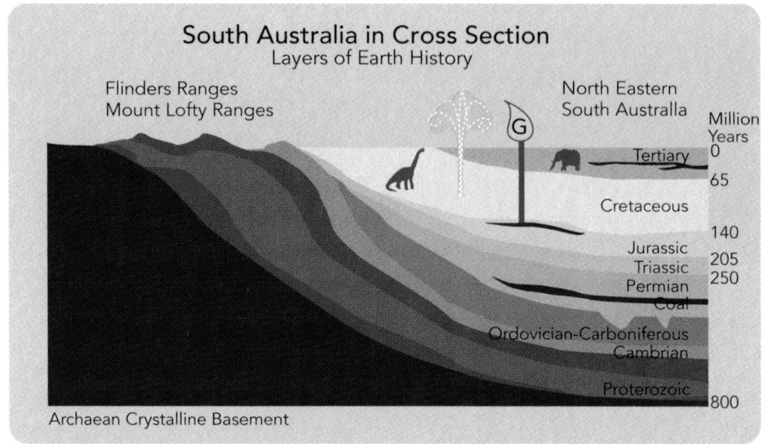

Above: Geologists study sedimentary rocks using surface outcrops, drill-holes, and seismic records to locate potential sources of petroleum and mineral resources. Top left: Power plants burn oil, coal, or natural gas to create steam that turns turbines on a generator to produce electricity.

Rigs pump crude oil (petroleum) from underground pools or reservoirs of porous rock. The oil is composed of complex hydrocarbons that need to be distilled into usable constituent fuels, such as gasoline, kerosene, and diesel.

burned in the presence of oxygen produce carbon dioxide, water vapor, and heat. However, the true reaction is rarely that simple or clean, as many of these compounds are impure in their natural forms. Some of the by-products released during combustion are toxic—sulfur dioxide, mercury, nitrogen oxide, and soot—and repeated or prolonged exposure to these chemicals can harm humans, animals, and the environment. For example, because coal contains sulfur impurities, the combustion of this fossil fuel produces sulfur dioxide, the primary cause of acid rain.

It is also worth keeping in mind that some fuels burn more cleanly than others. Methane, for example, is the chief component of natural gas and is made up of one carbon atom and four hydrogen atoms. It burns cleanly in comparison to coal and fuel oil because it produces more energy and less carbon dioxide, a historically benign substance that is now implicated in global warming.

Fossil fuels are nonrenewable resources. They take so long to manufacture naturally that it would be impossible or impractical to wait around another 300 million years while nature mixed up another "batch." That's why energy conservation is so important: When humans use up the world's reserves of fossil fuels, they will be gone forever.

GENERATING POWER

What happens in a huge electric plant depends on the type of fuel being burned. Every utility company uses a different mix, but coal is still the most popular. The coal is burned in huge chambers, and the resulting heat is used to boil water into steam, which in turn spins giant turbines that generate electricity. Sometimes powerful winds generated by the coal fires turn the turbines directly. It's ironic that at the heart of our fast-paced, digital society lies a very mechanical process that is not much different from the one that powered factories at the dawn of the Industrial Revolution.

Above: Power plants that burn coal emit millions of tons of sulfur dioxide, the primary cause of acid rain. Top: Unlike the mining operations of yesteryear, modern mines are carefully regulated to ensure that water surrounding the mine is protected from contamination.

Electricity

Since the dawn of time electrical currents have ripped through the sky, seeming to tear holes through clouds and, occasionally, sparking more than a little interest when they touched down to Earth. Lightning was one of the first signs to man of the existence of a special kind of force. As electric charges ionized the air, making bright streaks that were visible to the naked eye, curiosity was piqued about what that force was and how it could be harnessed.

No one person is responsible for giving shape and form to the ideas used to control and generate electricity. Many observations and inventions have been compounded over time. From the first time the ancient Greeks became aware of what we now refer to as static electricity, through Benjamin Franklin's escapades in a lightning storm armed with a kite and a key, and straight on through to the development of cathode rays and hydroelectric power, a lot of contributions have been made.

A FEW GOOD ELECTRONS

But what's most astounding is that all of this history, all of this invention and this genius, stems from some very tiny, very excited electrons. The electrons that spin furiously around the atoms that are the basis for all the matter that exists in the universe are negatively charged. When an atom is in balance, the negative charges of the electrons are balanced by the positive charges of the protons in the atom's nucleus. The electrons that are closest to the nucleus feel the strongest attraction and are held on to the tightest, so to speak, by the nucleus. However, the electrons in the shells or clouds farther away from the positive charge of the protons in the nucleus feel a slightly weaker pull from the nucleus. These electrons can be lured away and caused to jump ship to a neighboring atom. The energy contained in this flow of elec-

Above: Lightning, very much like static electricity, occurs as the result of the attraction between oppositely charged particles. A bolt of cloud-to-ground lightning is an exchange of electricity between charged regions in a cloud and oppositely charged regions on the ground. Top left: The molecules and atoms within the hot stove burner begin to move and oscillate more quickly as the temperature increases.

trons is called electricity. One of the reasons that metals are such good conductors of electricity is that they fall into a class of elements that have a particularly easy time giving up electrons, as they tend to hold on to them more loosely.

Benjamin Franklin's legendary kite experiment successfully showed that lightning is actually static electricity and led to the invention of the lightning rod. Although there are various accounts of this experiment, the most prevalent version maintains that Franklin attached a metal spike to a kite, which he then flew during an electrical storm. Lightning struck the kite and electricity traveled down the kite's cord to a key tied near the end, creating a spark when Franklin brought his knuckle close to it.

GETTING IN THE FLOW

"Opposites attract . . .We have such chemistry . . . There's a real spark between us . . ."

Much like an ill-fated romance, electricity involves two opposite entities that are uncontrollably drawn to each other. This is rather unromantically referred to as the First Law of Electrostatics: Unlike charges attract each other, like charges repel each other. It is this undying attraction and repulsion that keeps the transfer of electrons moving. Once an atom has lost an electron, it has a positive charge because there is now one less negative charge to balance out the positive charge of the proton. This positively charged core (called an "ion") is now anxious to get back into balance and will draw in another negative charge—in the form of a free electron—and so the chain continues. This passing of electrons, like a subatomic game of hot-potato, helps clarify the idea of flow. Alessandro Volta, namesake of the volt, proved that moisture between two dissimilar metals could generate an electrical current and in doing so showed the world not only that electricity existed but that it could "flow": It could travel from one place to another along an electric wire. His "voltaic pile," essentially the first electric battery, was composed of moist chemicals sandwiched between two different metals. This took investigations into electricity in very important directions.

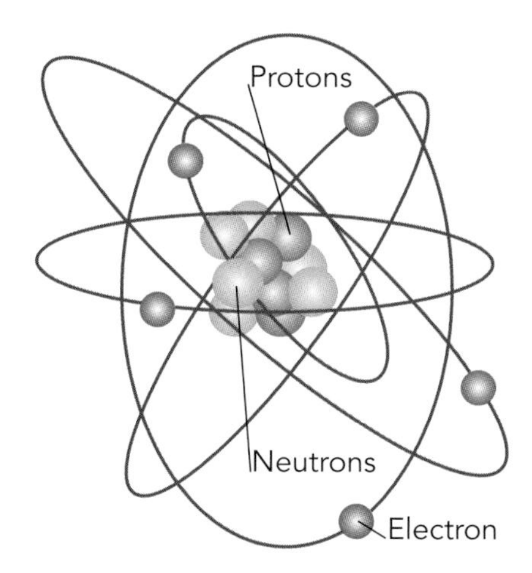

When an atom is neutral, the number of positively charged protons (pink) and negatively charged electrons (blue) are balanced. When the atom loses an electron, it takes on a positive charge and is referred to as an ion. In this unbalanced state, the positively charged ion will be attracted to a negatively charged ion.

Electricity is referred to as a secondary energy source. In other words, some other sort of energy is converted to give us electricity, whether it's energy from the Sun or energy from the combustion of fossil fuels. Electrical energy can then be transported to where it is needed, then converted into new forms to do useful work.

Using Electricity

The flick of a light switch is such a simple gesture, one performed more times a day around the world than one could imagine. A lot happens in the flick of a switch, tapping into centuries or experimentation into how to generate, control, and transport electricity.

Above: Developed during the second half of the nineteenth century, steam turbines extract thermal energy from steam and use this energy to generate electricity. Top left: When you flip on a light switch, you are creating a circuit that electrons can travel along to deliver electricity. Flipping the switch into the off position breaks the circuit and stops the flow of electrons.

ELECTRICITY ON THE GO

Batteries today generate electricity much in the same way that Volta's voltaic pile did at the beginning of the nineteenth century. A battery contains chemicals—an electrolyte—and two different types of metals, one at either end of the battery. The reaction between the chemicals and the metals causes electrons to free themselves from their atoms, but this happens more readily with one of the metals than the other. The result is that one end of the battery, the anode, is more negatively charged than the other end, the cathode. The free electrons in the anode want desperately to get to the more positive cathode end of the battery, but the chemicals between them prevent that from happening. When a wire is connected to both ends of the battery, the electrons can race along, hopping from atom to atom toward the cathode, trying to balance the charge. This completes a circuit in which the electrons can travel and creates current. Along the way, that current can do work, like power a radio or a flashlight.

A battery converts chemical energy into electricity, causing electrons to flow from the negative to the positive terminal. Inserting a battery into an electrical appliance connects the terminals and creates a circuit through which the electrons can flow.

POWER TO THE PEOPLE

Electricity is secondary energy, meaning that the key function of a power station is to convert mechanical energy or chemical energy into electric energy. A common way this mass production of electricity is accomplished is by using knowledge of electromagnetic induction. Thanks to Michael Faraday—and several scientists whose prior work supported his experimentation—the relationship between magnetism and electricity is better understood. Faraday's law of electromagnetic induction, which he published in the 1830s, explains that

At an electrical substation, voltage is transformed—often from high industrial voltage to a lower voltage suitable for consumer use. Ceramic insulators, such as those seen above, isolate high-voltage lines from the ground.

Portable generators contain combustion engines that convert fuel into energy.

WHEN THE LIGHTS GO OUT

One doesn't really appreciate something until it is lost. The modern world's dependence on electricity becomes terrifyingly clear during a power outage. Standby electric generators have long been used as backup power supplies for hospitals and city government offices. Now, in increasing numbers, they are being used by individual homeowners as well. Generators work like the engine of a car. Combustion is the source of energy—fuels like gasoline, natural gas, or propane are used—but instead of using this energy to spin axles, the generator spins a magnet inside a stationary coil of wire. As the poles of the magnet swing past the coil, the changing magnetic field generates a changing current in the wire. This alternating current (AC) can be plugged into your home and used as a substitute for the electricity grid, shedding new light on a difficult situation.

a conductor in a changing magnetic field can produce an electrical potential difference, also referred to as voltage. So, when a material capable of conducting electricity moves across a magnetic field—movement is key—an electric current can be induced in that material.

The majority of electricity generated in the United States is generated by steam turbines. Steam causes the blades of a turbine in a generator to spin, rotating the turbine's shaft. This rotating shaft is connected to a coil of wire, which spins inside a stationary magnet. As the wire coil rotates, its movement in relation to the existing magnetic field produces an electric current in the wire. This current flows from the generator to begin its journey to homes and business miles away. Transformers, developed by William Stanley, allow the voltage to be "stepped up" for efficient travel over long distances and then "stepped down" again for local distribution.

Of course, to make the steam that spins the blades that rotate the shaft, coal or other fuel is burned and used to heat water, which then creates the steam. This multistep process involves an investment of energy along the way and, overall, is not the most efficient method for producing electricity. It is estimated that the amount of energy spent generating the electricity—the burning of fossil fuels, the mechanical energy used—is three times as great as the amount of electricity actually produced. Still, more than half the electricity generated in the United States is from power plants run on coal, which is a nonrenewable resource. The limited supply of fossil fuels, the resulting pollution, and the limited electrical energy returned in relation to the amount of energy invested have created an ever-increasing need to use alternative means of generating electricity, such as solar and wind power.

Alternative Fuels

In recent years, scientists, political leaders, and others have become more interested in identifying and perfecting the use of renewable energy resources. Unlike fossil fuels, which will one day run out, renewable energy is tantalizing because there is an inexhaustible supply of it. Once humankind perfects a way to use sunlight, wind, water, and plant products to run its machines and cities, even the poorest nations could enjoy the benefits of prosperity, because after the initial investment in equipment is paid off, the energy would be forever free.

SOLAR POWER

Capturing sunlight for electricity is accomplished by means of solar, or photovoltaic panels, large sheets of glass-encased solar cells that absorb the Sun's rays. Solar cells are made of two layers of silicon—one containing small amounts of phosphorus impurities, the other with small amounts of boron. When photons, particles of sunlight, strike near the boundary between the layers, they excite electrons that flow to the phosphorus-laden top layer of the panel, imbuing it with a negative charge. As electrons flee to this upper layer, the bottom layer is left positively charged. The panel, in a sense, behaves like a giant battery, charged up by sunlight.

The resulting current is routed directly into the electrical wiring of a building. Excess electricity can be stored in large battery banks on the premises, or sent to one's local power utility to be distributed as electricity to other paying customers. This "grid intertie" technology, already in use in many homes, means that the average home has the ability to become a mini–power plant, selling its excess power to local energy providers.

WIND POWER

Most people are familiar with the scene of a windmill spinning constantly in the countryside. Perhaps they vaguely understood that the blades of the windmill, propelled by gusts, were actually performing real work, such as operating a water pump. Modern wind power uses similar technology. As the blades turn, they spin a strong steel shaft that operates an electric generator, usually mounted in an enclosed compartment directly behind the blades.

Above: BedZED, which stands for "Beddington Zero (fossil) Energy Development," is an environmentally friendly and energy-efficient eco-village in south London that opened in 2002. All energy used in BedZED is generated on-site using renewable sources. Top left: Photovoltaic (PV) cells are a familiar solar technology used in a wide range of products, including solar-powered garden lights, watches, and calculators. Made of semiconductors, such as silicon, PV cells can convert light into electricity without making noise or causing pollution.

The most common generators manufacture electricity by spinning magnets to generate an electromagnetic field. This field can be tapped and sent from the windmill to large power plants, small home battery storage cabinets, or directly to some indoor appliance or machine. Despite its promise, wind power has become controversial in recent years because many communities object to the construction of wind turbines, claiming they make the landscape less attractive. Currently wind power supplies only 1 percent of the world's power. But some nations have fast-tracked this form of energy; 25 percent of Denmark's electricity comes from wind.

A test car is filled with liquid hydrogen fuel at a solar hydrogen filling station in Neunburg vorm Wald, Germany. The white storage tank (far right) contains hydrogen fuel derived from water using a process called electrolysis and power from photovoltaic solar cells (far left).

BIOFUELS

In 1900 a man named Rudolf Diesel (1858–1913) debuted a new type of engine at the World Exposition in Paris. Instead of requiring a spark to burn the fuel, his engine ignited a fuel made from peanut oil solely through strong pressure. Vegetable-based fuels would have been beneficial then as now, since users would be able to grow their own fuel instead of relying on outside imports or the costly refining of petroleum. Any modern Diesel engine will burn vegetable oil, a fact that is being widely explored now that the world looks to alternative sources of fuel. Some diehard fans of the technology go so far as to convert used vegetable grease—straight from a restaurant fryer—in their garages and use it as fuel. Others see promise in ethanol, the grain-derived chemical compound also known as ethyl alcohol, found in beverages such as whiskey. However, a growing body of research suggests that it may require more fossil fuel energy to grow, fertilize, and process corn into ethanol than the energy you get from burning the ethanol. In the United States, "flex-fuel" vehicles have already been built to accept ethanol as a fuel, but the product is a long way from being widely available. The nation's fuel delivery network of gas stations would need to be completely overhauled to accommodate ethanol.

Today, the price of wind power is comparable to that of coal, and in Europe, the capacity to generate electricity using wind turbines is growing at about 10 percent per year.

CHAPTER 6

INSIDE THE ATOM

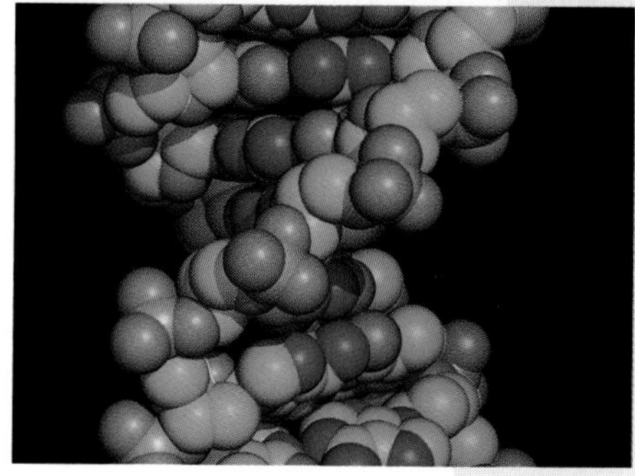

Left: Like a miniature solar system, an atom is composed of negatively charged electrons orbiting a nucleus of protons (positively charged) and neutrons (which have no charge). Top: The mushroom cloud released by a detonating hydrogen bomb has become an enduring symbol of the massive destruction of successfully manipulated atomic energy. This is an image of an unidentified U.S. nuclear weapons test in the 1950s. Bottom: The now-familiar double helix of a DNA strand. Each sphere represents a single atom.

The atom is a universe unto itself: a center surrounded by orbiting particles, like an infinitesimal sun flanked by electronically charged planets. Combining and splitting, atoms and their constituents are constantly taking on new and different characteristics. For human beings, understanding the atom and all its moods was the first key to understanding Earth and the cosmos beyond it. The spontaneous decay of some unstable atoms gives us a timeline, etched in the minerals and fossils from billions of years ago, that allows us to better understand the life of our planet. Manipulating the nucleus of an atom has resulted in radioactive pharmaceuticals that can be followed like a guiding light to the site of disease in a hopeful patient, as well as the hope of new sources of energy. The elements found in our hearts and brains got their start in the stars of the great beyond. But with knowledge has also come great responsibility. The same atoms that form the basis of all the life in the universe can also be manipulated to extinguish it.

Isotopes

Every element has a specific number of protons, neutrons, and electrons and is associated with a particular atomic mass. But elements can take on different forms to suit a particular occasion. Atoms gain and lose subatomic particles in order to serve a particular purpose or participate in a particular reaction. Generally speaking, isotopes are like different-size versions of the same element, but this difference gives the isotope properties that make it different, and these differences have valuable applications.

The Nuclei of the Three Isotopes of Hydrogen

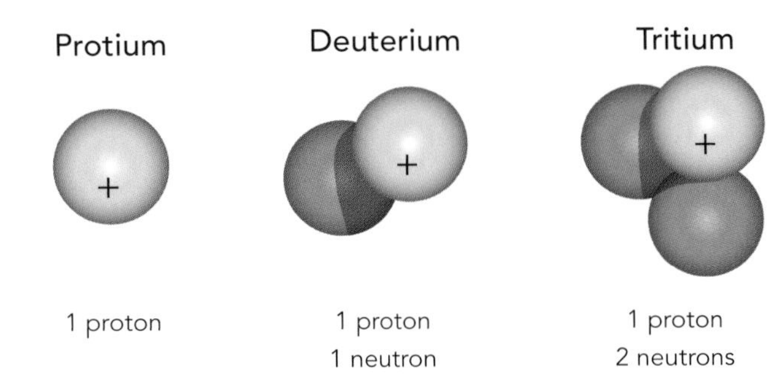

Isotopes of an element contain different numbers of neutrons but the same number of protons. The three hydrogen isotopes can have one, two, or no neutrons coupled with a single proton.

THE SAME BUT DIFFERENT

An isotope of a particular element is still that element. However, it is a form of that element that contains a different number of neutrons. Elements are identified by their atomic number, or the number of protons they have. So when a particular element loses or gains a neutron, it does not change what that element is. But one way in which the formation of an

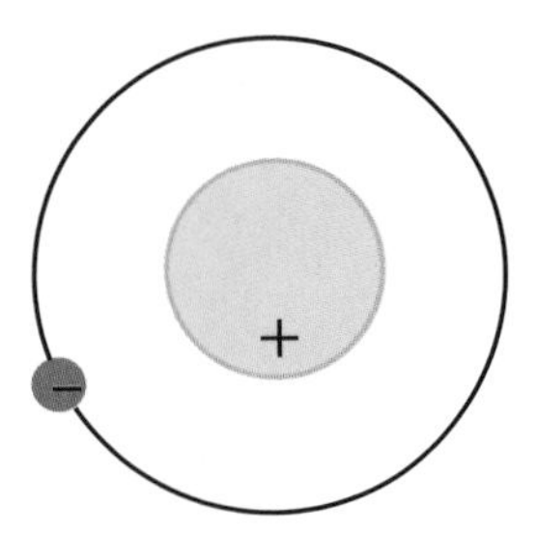

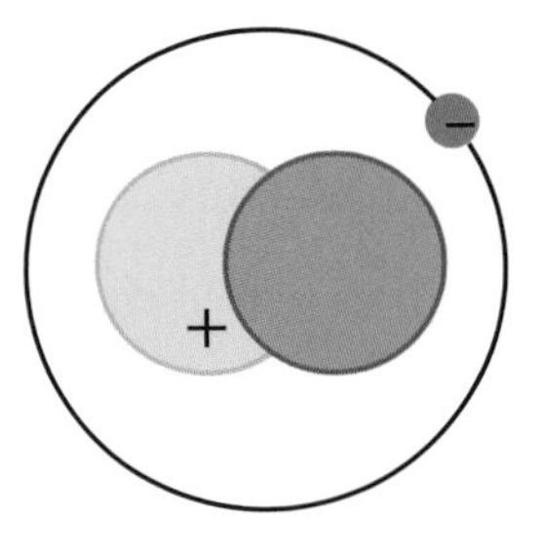

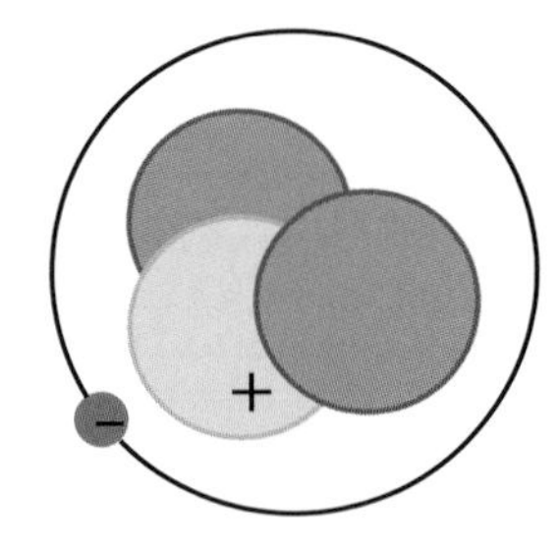

Above: An element's weight is altered by the addition of neutrons. Thus isotopes have different atomic masses, as shown by the comparison of hydrogen and its isotopes. Top left: The age of this small fossil embedded in rock can be determined with the help of radioactive isotopes.

isotope does change the characteristics of an element is its effect on atomic mass. When an element loses or gains electrons, creating an ion, the weight of the atom is hardly affected. But gaining or losing bulkier neutrons is a different story.

Because each isotope of a particular element has a different atomic mass, the atomic mass that is seen on the periodic table actually represents an average of the masses of the different isotopes that exist for that element.

Carbon 14 is an isotope that is fairly familiar to most people because of its use in carbon dating, which is discussed later in the chapter. The "14" here represents the number of nucleons—protons and neutrons—in this particular isotope of carbon. Taking a glance at a periodic table shows the atomic number and atomic mass of carbon to be 6 and 12 respectively. That means that this particular isotope of carbon has two more neutrons than usual.

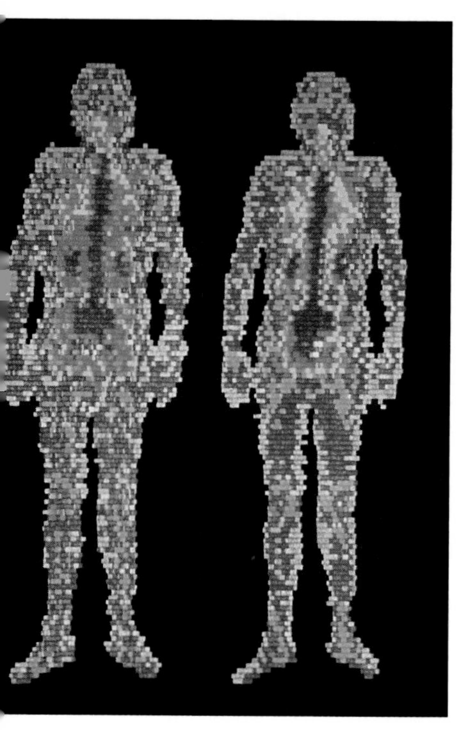

Radioactive isotopes, injected into the human body, can help identify medical problems. This man has been injected with an isotope attracted to bones, and thus his skeletal structure, especially his spine and pelvis, are highlighted.

Chemically, different isotopes of the same element generally behave the same, though heavier isotopes tend to move along more slowly in reactions than the other isotopes. Many isotopes are completely stable. However, things do not always remain so calm deep inside the nucleus. Neutrons can help to stabilize a nucleus by separating the protons, cutting down on the "like repels like" electrostatic forces that would tend to hurl the protons apart. Too many neutrons can also destabilize a nucleus, however, leading it to fall apart into lighter pieces. The ratio of protons to neutrons can affect the stability of the nucleus, and if it is too high or too low for a particular isotope, the isotope may begin to decay. When this happens, the isotope is radioactive, also called a radioisotope.

ISOTOPES IN ACTION

Though many isotopes of many elements do exist naturally, new ones can also be produced in a lab by firing neutrons at an element. Isotopes and radioisotopes serve many different purposes in nuclear chemistry, from nuclear medicine to nuclear power. Stable or unstable, isotopes are easily identifiable and can be distinguished from other atoms of the same element. Because isotopes have different atomic masses, mass spectroscopy can distinguish the various isotopes of a given chemical element. This makes them ideal markers or tracers in chemical reactions. This technique is called isotopic labeling. It's as though in a sea of chlorine atoms, for example, one particular atom is wearing a pair of flashy red pumps—it stands out. One example of how a stable isotope might be used is in the field of pharmaceuticals, to help study how a particular drug is metabolized. The ability to perform various types of spectroscopy (which are discussed further in chapter 11) also depends on the properties of different isotopes, as does the ability to date fossils and estimate the age of the Earth.

A scientist for the U.S. Geological Survey collects gas samples for carbon 14 analysis. The analysis will be applied in this case to study the flow of gases through different environments.

Radioactivity

Radioactivity is probably one of the most misunderstood concepts of the nuclear age. Thanks to Hollywood, we have been inculcated with the notion that a dose of radioactivity will, among other things, breed gigantic lizards bent on destroying major cities, transform meek adolescents into web-slinging superheroes, or smite us all dead. The prosaic truth is far more fascinating: Some atoms are always disintegrating before our eyes, and this is often a benign reaction.

Radioactivity refers to the tendency of certain, inherently unstable elements to degrade. When they do, they emit particles from their nuclei without notice. When this happens, the original nucleus is forever altered. This random, natural occurrence is called a *decay* event, and the element is said to have undergone radioactive decay.

Radiation, leftover heat from the big bang, is non-ionizing, so relatively harmless. It is all around us, springing from rocks in the Earth and showering us from space. Our dinner plates are radioactive, thanks to the Earth-derived elements used to make and decorate them. Our food is radioactive, coaxed as it was from the Earth. When you fly in an airplane, above the protection of the Earth's atmosphere, you are exposed to even more cosmic radiation from space. Why, even an innocent glass of milk will light up a Geiger counter. All life on Earth has evolved with radioactivity in mind. Our genetic code is well aware of its presence, and has made adjustments over millennia to accommodate safe amounts of it.

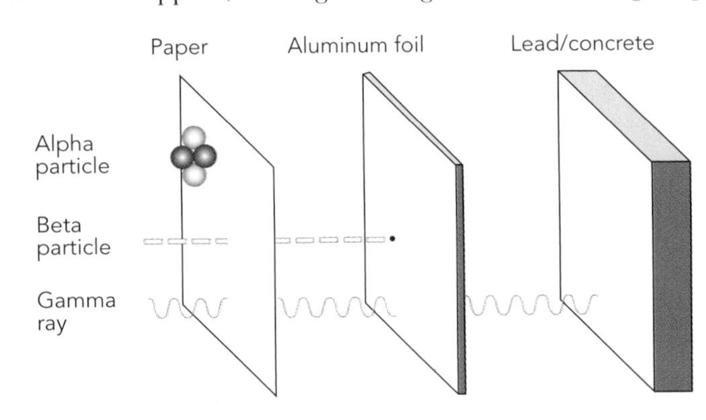

Above: Of the three particle types released by a splitting nucleus, the heaviest, or alpha particle, can be stopped with paper, a beta particle can be stopped with aluminum, and it takes a block of lead to halt a gamma ray. Top left: The requisite warning on containers of dangerous radioactive material has helped color public conception of radioactivity as harmful, or even frightening.

Radioactivity was first noticed in 1896 by Antoine-Henri Becquerel (1852–1908), a French physicist, who was puzzled to find that an unknown source of radiation was blackening the photosensitive materials in an experiment he was conducting. Numerous other scientists analyzed this mysterious source, most prominently Ernest Rutherford (1871–1937), the so-called father of nuclear physics, who named the nucleus and its components, and observed that when a nucleus spontaneously split, it released three major types of particles, which he named *alpha, beta*, and *gamma.*

Alpha particles consist of two protons and two neutrons, and are essentially helium atoms. They are positively charged, travel slowly, and are so heavy that they can be stopped in their flight by a sheet of paper. Betas are lighter, and fleet of foot. Only about 1/2,000th the mass of a proton, they are high-energy electrons. Beta particles are negatively charged and can be stopped by a plate of aluminum.

Gammas behave as both particles and waves. They are neutral and can travel through most substances, although they can be stopped by a thick slab of lead.

Any given radioisotope will disintegrate in rather predictable ways. Each is associated with a length of time called its "half-life"—the amount of time it takes for half of the original atoms to decay. If the half-life of an element is three hours, this means that in three hours, only half of the material will be left, the rest having been converted into other materials. In six hours only one-quarter of the original atoms will remain, and so on—the amount of radioisotope continually decreasing, though never quite disappearing.

USEFULNESS OF RADIOACTIVITY

The discovery of radioactivity forever altered human life on Earth. Once we learned to manipulate the atom and use the offshoots of certain elements—such as uranium and plutonium—to penetrate living and inanimate matter, humans at once assumed a greater burden of responsibility for the planet.

On one hand, radioactivity can be used for good. We use it today in such fields as medicine, manufacturing, security, food production and sterilization, and the sciences. On the other hand, excessive amounts of radioactivity can also damage the human body, wreaking havoc with our genetic material beyond our ability to heal. Alpha particles, in particular, can strip electrons from atoms in living cells, programming them to impart misleading information to human DNA. That is why excessive amounts of radioactive materials must always be handled with extreme care.

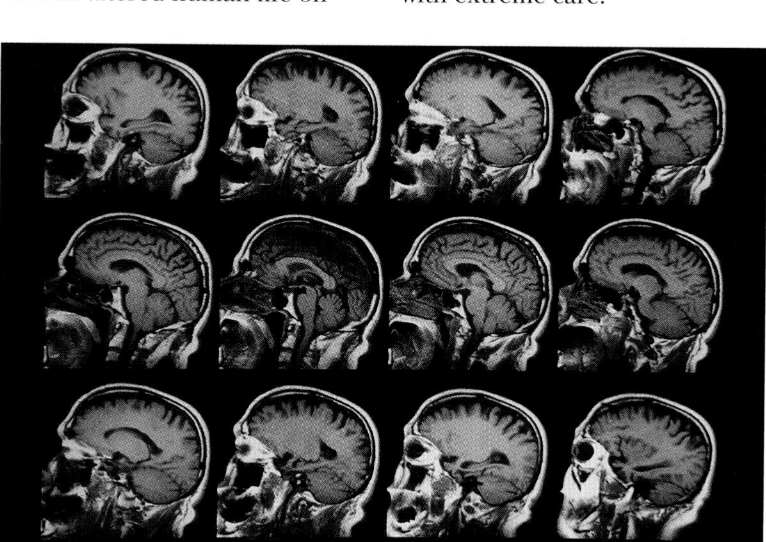

X-rays and other imaging techniques are examples of the ways in which understanding atoms, isotopes, and how they function has improved our ability to diagnose and cure diseases and injuries.

George Gamow, a pioneering contributor to the big bang theory. First developed in 1948, the theory has yet to be replaced by scientists.

AS EASY AS ALPHA, BETA, GAMMA

One of the most important scientific concepts of all time, the big bang theory, is also linked to a famous practical joke. The scientific paper that first theorized that the universe originated in a dense, hot state was written in 1948 by a young American doctoral student named Ralph Alpher. His thesis supervisor and co-theorist was a Russian physicist named George Gamow. When Alpher completed his paper, Gamow playfully suggested that they share the writing credit with yet another scientist, Hans Bethe, who had not done any work on the theory. An inveterate prankster, Gamow knew that the published paper's byline would read "Alpher—Bethe—Gamow," a play on words that simultaneously invoked the three famed particles of radioactive decay—*alpha, beta, gamma,* named for the first three letters of the Greek alphabet—and hinted at the beginning of all things.

Nuclear Energy

Nuclear power plants, with their tall stacks forever spewing steam, have become controversial icons of the modern world. Nuclear energy is simultaneously hailed and condemned, lauded for its ability to generate copious amounts of power and denigrated for the awesome burden of its radioactive waste. The science behind the technology dates to World War II, when a small cadre of physicists led by J. Robert Oppenheimer (1904–67) began work on the Manhattan Project, determined to build the world's first nuclear weapons. The team's work ultimately led to the creation and detonation of three bombs: a test explosion over New Mexico, and the infamous bombs dropped over Hiroshima and Nagasaki in August 1945. On the more peaceful side, the Manhattan Project led to the first controlled self-sustaining nuclear chain reaction (forerunner of a modern reactor), performed by Enrico Fermi and Leo Szilard at the University of Chicago in 1942.

Following the war, nuclear science was exploited by many nations to create an effective means for generating electricity. Scientists discovered that when a single atom of enriched uranium (U-235) is bombarded with one neutron, the result is the production of 10 million times more heat than is produced by burning a single carbon atom in a lump of coal. This reaction is called nuclear *fission,* because the uranium atom splits, producing energy, smaller nuclei, and neutrons, which may split other uranium nuclei to produce still more energy. In modern nuclear power plants, a controlled chain reaction of this sort is used to generate heat, which boils water, producing steam to turn turbines. Nuclear fission is different from nuclear *fusion,* the means by which the Sun manufactures energy, by fusing four hydrogen atoms into a single helium atom.

Above: Mushroom cloud from the bomb dropped on Nagasaki, Japan, on August 9, 1945, as viewed from a bomber. Nagasaki was the second—and last—target of a nuclear weapon, and its destruction is evidence of the incredible power of nuclear energy. Top left: Smokestacks spewing steam are a familiar site at nuclear power plants around the world.

THE ANSWER TO GLOBAL WARMING?

Nuclear power plants produce only 7 percent of the world's energy, but their importance varies widely among nations. France, for example, derives nearly 80 percent of its energy from nuclear power plants; Japan 30 percent, the U.S. 20 percent. There are about 557 reactors in the world, and about 400 of them are operational. Following nuclear accidents at Three Mile Island in the U.S. (1979) and Chernobyl in the former Soviet Union (1986), public confidence in of nuclear energy

The impact of the Chernobyl disaster in 1986 continues today, having not only direct repercussions for its survivors but influencing the status of power plants and the use of nuclear energy worldwide. It is a reminder of the destruction even well-intentioned use of nuclear power can produce.

has been somewhat diminished, and several nations have voted to end their nuclear programs. But increasing concern about the possibility of global warming has led to renewed interest in nuclear power, which does not produce carbon dioxide, one of the so-called greenhouse gases believed responsible for rising world temperatures.

The chief objection to nuclear energy is the possibility of accidents in which radioactivity could be dispersed in the atmosphere. In nuclear power plants, small pellets of uranium fuel are contained inside steel rods, which, when engaged in a fission reaction, give off tremendous heat to boil water. Every couple of years, the used fuel rods must be switched with fresh ones. Though spent, the fuel rods continue to emanate radioactivity and must be stored away from humans to avoid contamination. Following the internal chronology of radioactive half-lives, after only 10 years the rods are 1,000 times less radioactive. After 500 years they are less radioactive than the ores—coal, granite, and the like—from which they were originally derived. But some fuels produce rods that will be dangerously radioactive for up to 100,000 years.

Storing, handling, and managing this flow of waste is a tremendous burden to place not only on our society, but on successive generations. Ideally, the waste should be stored in a place far from human activity. Various nations are struggling with this issue. Following decades of scientific study and political debate, the U.S. had planned to store its waste in a single underground repository at Yucca Mountain, Nevada, but this plan has been held up by numerous objections and lawsuits. Proponents of nuclear power say its benefits outweigh the risk, and that while expensive, nuclear power is still far cheaper than other forms of alternative energy. Detractors say that the chance of disaster is too great to ever make nuclear power acceptable, and they argue that this labor-intensive form of energy is simply not cost-effective.

Radiometric Dating

Atoms can behave a lot like people: In general, they seek stability. And while circumstances may arise that result in an unstable situation, the desire is always there to seek a more stable state. While many isotopes are entirely stable, there are those that are not. Carbon 13 is relatively stable, but if one more neutron is present in the nucleus the result is carbon 14, which is radioactive. In the case of uranium, however, all isotopes of this element are radioactive. These more tempestuous versions of elements cannot exist in this state forever, and so they decay. But not only do they decay—they do so at very specific, measurable rates.

The process of using these rates of decay to determine the age of materials like rocks, minerals, and organic fossils that contain these radioisotopes is called radiometric dating. These naturally occurring radioisotopes have been in the bones of the Earth since it was formed. Seeking stability, these radioactive "parents" decay into more stable "daughters." The number of parent atoms that are present in a sample combined with the number of daughter atoms present gives the number of atoms that were originally present in the material. Once this is known, then the number of parent atoms originally present and the number of parent atoms still existing can be used to determine the material's approximate age. The half-life of the radioisotope—the amount of time it takes for half of a given amount to decay—is a constant that can be applied to the ratio between the number of the original parent atoms and how many are still left.

One familiar method of radiometric dating is radiocarbon dating, which uses the half-life of carbon 14 to determine the age of organic plant and animal fossils. As carbon 14 decays, it eventually becomes nitrogen 14. Carbon 14 has a half-life of approximately 5,730 years. Because of the length of its half-life, radiocarbon dating

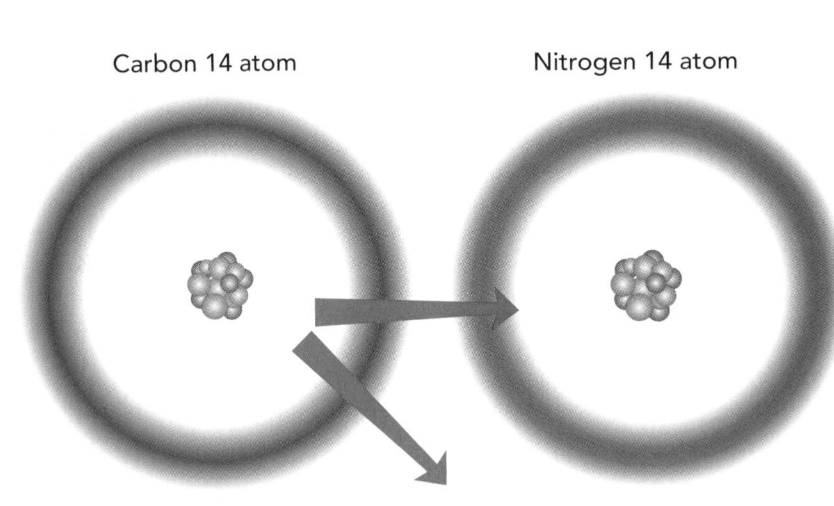

Above: Carbon 14 decays steadily into a nitrogen 14 atom, releasing beta particles as it does so. Because of its presence in organic life forms, carbon 14 can be used to date fossils up to 50,000 years old. Top left: Fossils that are more than 50,000 years old can be dated using the regular rate of decay of radioactive isotopes such as uranium 235 or potassium 40.

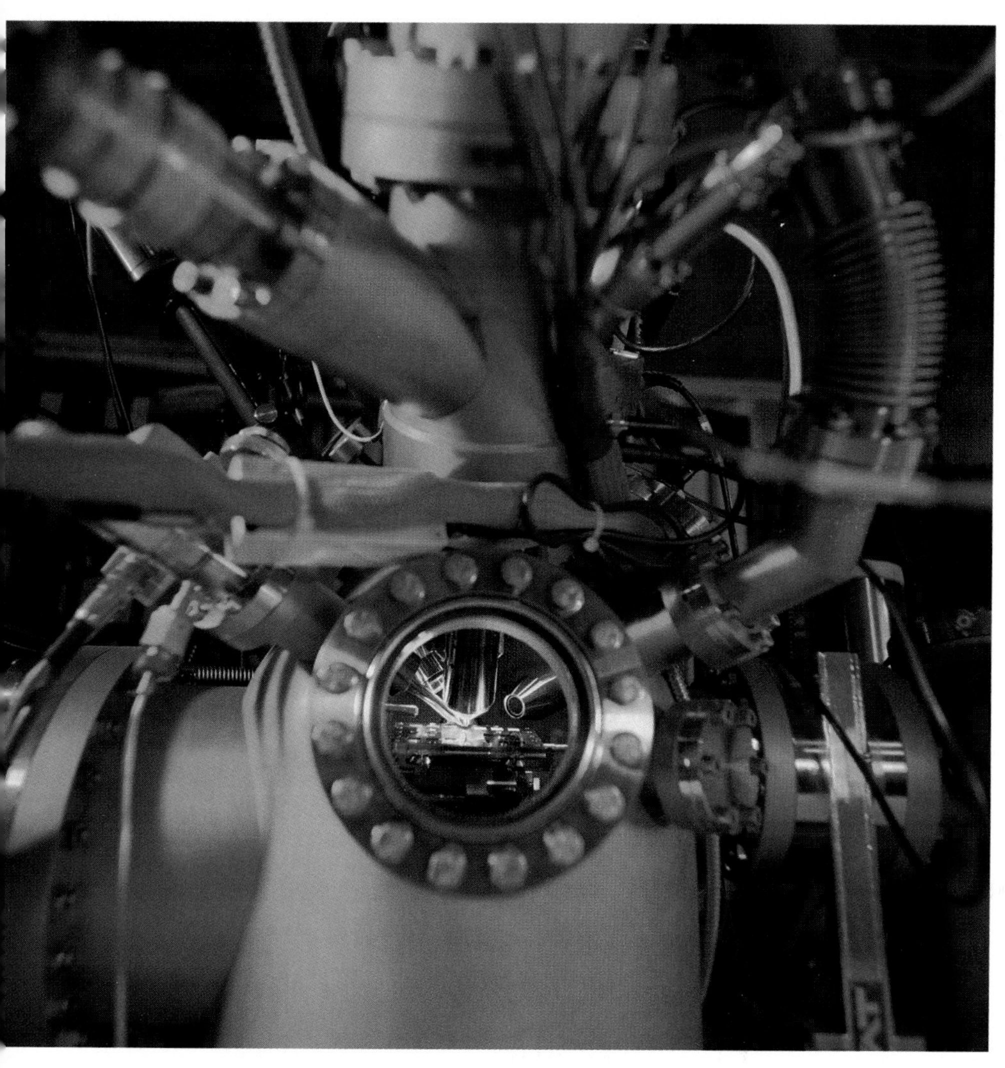

This mass spectrometer determines the chemical composition of a sample by converting particles to ions and measuring their mass. Sophisticated machinery like this has enabled scientists to discover much about the world and its past.

cannot be used on fossils more than approximately 50,000 or so years old, because there would be such a minute amount of carbon 14 left in the organism or plant. The shorter the half-life, the more limited the range that the radioisotope can date. Fifty thousand years sounds like a long time, but when scientists are trying to determine the age of a volcano that may have rained lava down upon a stegosaurus during the Mesozoic era, then radiocarbon dating is hardly adequate.

Uranium's half-life is another story. Two common radioisotopes of uranium can be used to determine the age of materials billions of years old. Uranium-lead radiometric dating is so named because the uranium decays into lead. Uranium 238 decays to lead 206 and has a half-life of a whopping 4.47 billion years. Uranium 235 decays into lead 207 and has a half-life of a respectable 704 million years. As these isotopes decay into lead, measuring the proportion of lead to uranium in a specific sample of a rock or mineral allows for the estimation of the age of the material in which is was found.

Radiometric dating has been used to determine the age of Earth, which is estimated to be approximately 4.6 billion years old. Beyond the dating of individual fossils, rocks, minerals, and planets, radiometric dating techniques have also been used to estimate the time at which major geologic events have occurred. Such methods have also helped determine the timing of the activities of glaciers and volcanoes. Mass spectrometers are sometimes used in conjunction with radiometric dating, making it possible to measure much smaller amounts of radioisotopes.

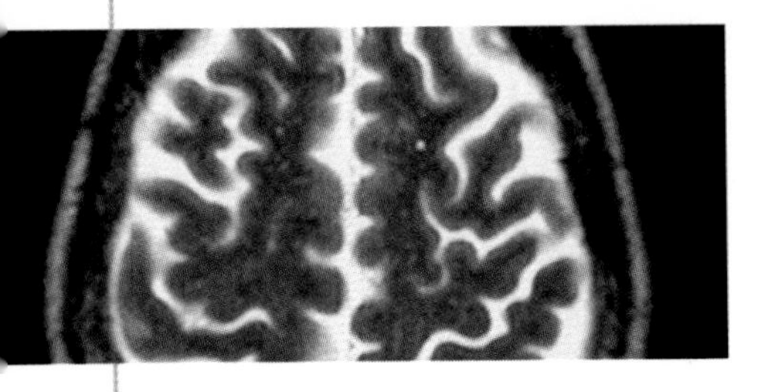

Nuclear Medicine

One of the most frustrating things for physicians is having a sick patient whose ailment is a mystery. Under those circumstances, peeking inside the human body is desirable, but no doctor wants to put patients through the unnecessary stress of exploratory surgery. That is why modern medicine employs noninvasive imaging techniques, such as X-rays, MRIs, CAT scans, and ultrasound. But advanced as these technologies are, they often do not provide enough detailed information. The resolution, or graininess, of the images may still be too coarse to provide good detail about what is happening inside a specific body part, such as a single kidney. In these cases, doctors may turn to the world of nuclear medicine.

In this field, doctors use radioactive drugs as "tracers" to help view the inside of a patient's body. In positron emission tomography, also known as a PET scan, the patient swallows, or is injected with, a drug that flows to or is absorbed by the problem area of the body. The drug contains radionuclides, lab-made atoms with unstable nuclei that are destined to have short half-lives. As these so-called radiopharmaceuticals decay, they emit positrons (anti-electrons) that interact with electrons in the cells of the body and shoot gamma rays out of the patient's body. Computers then reconstruct a three-dimensional image of the distribution of the radionuclide based on the distribution of these gamma rays.

WITNESSING THE PROCESSES OF LIFE

These rays are detected by a gamma ray camera, interpreted by a computer, and assembled into a three-dimensional image that the doctor can view easily. Doctors can see tumors, witness blockages causing poor blood flow, and even spot defective blood vessels. Tracers work well because certain ailments, glands, and organs are naturally attracted to specific elements. For example, tumors are attracted to phosphates. Inject an isotope of phosphorus near the site of a suspected tumor, and it will be sucked into the malignant zone and reveal the "hot spot." In a heart patient, a doctor can administer an unstable isotope of thallium, and, as the drug emits its radioactive signature, the doctor can chart the drug's movement through the blood vessels all the way to the heart.

There are some drawbacks to this technology. PET scans are expensive and not always available in all areas. The short lifespan of the drug means the imaging facility must be located near a particle accelerator where the isotopes are manufactured. Other types of tests are less costly because they use longer-lived isotopes and don't need to be in

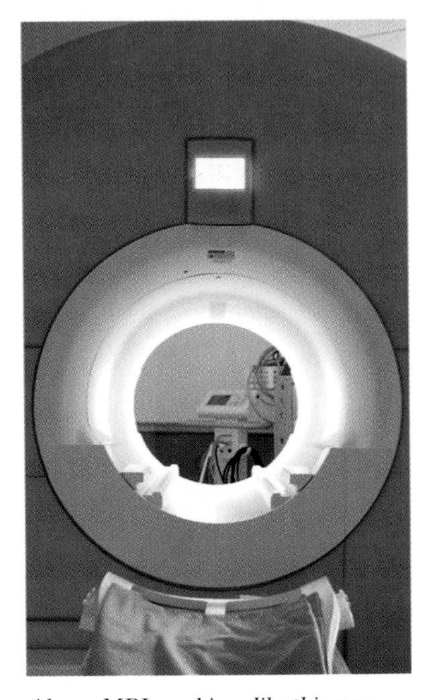

Above: MRI machines like this one use highly sophisticated technology to provide internal images of the body, but are still imperfect. Doctors may choose instead to use radioactive particles to identify internal problems. Top left: An MRI/MRA of the brain vasculature. Significant advances in brain imaging have been made in the past decade, and scientists continue to develop new techniques.

close proximity to the nuclear manufacturing facility.

SAFETY IN THE NEW MEDICINE

How safe is it to ingest radioactive medication? Do the benefits of the treatment outweigh the risks? Scientists and physicians agree that little danger is posed by the isotopes consumed or injected in the patient's body. The elements are radioactive only for a short time—a few minutes, hours, or a day—and then they break down naturally or are eliminated as waste by the patient's body. Indeed, we humans are bombarded by radiation every day that is far more powerful than those emitted by radiopharmaceuticals.

However, the medical community continues to debate the potential harm done by radioactive rays, such as X-rays or gamma rays, in the course of medical diagnoses. Yes, these rays can have a cumulative, harmful effect if used repeatedly, but in actual practice these diagnostic techniques are not employed that often on the same patient. Even if the doctor wanted to do so, the patient's insurance carrier would no doubt limit the usage due to the high cost of the procedures.

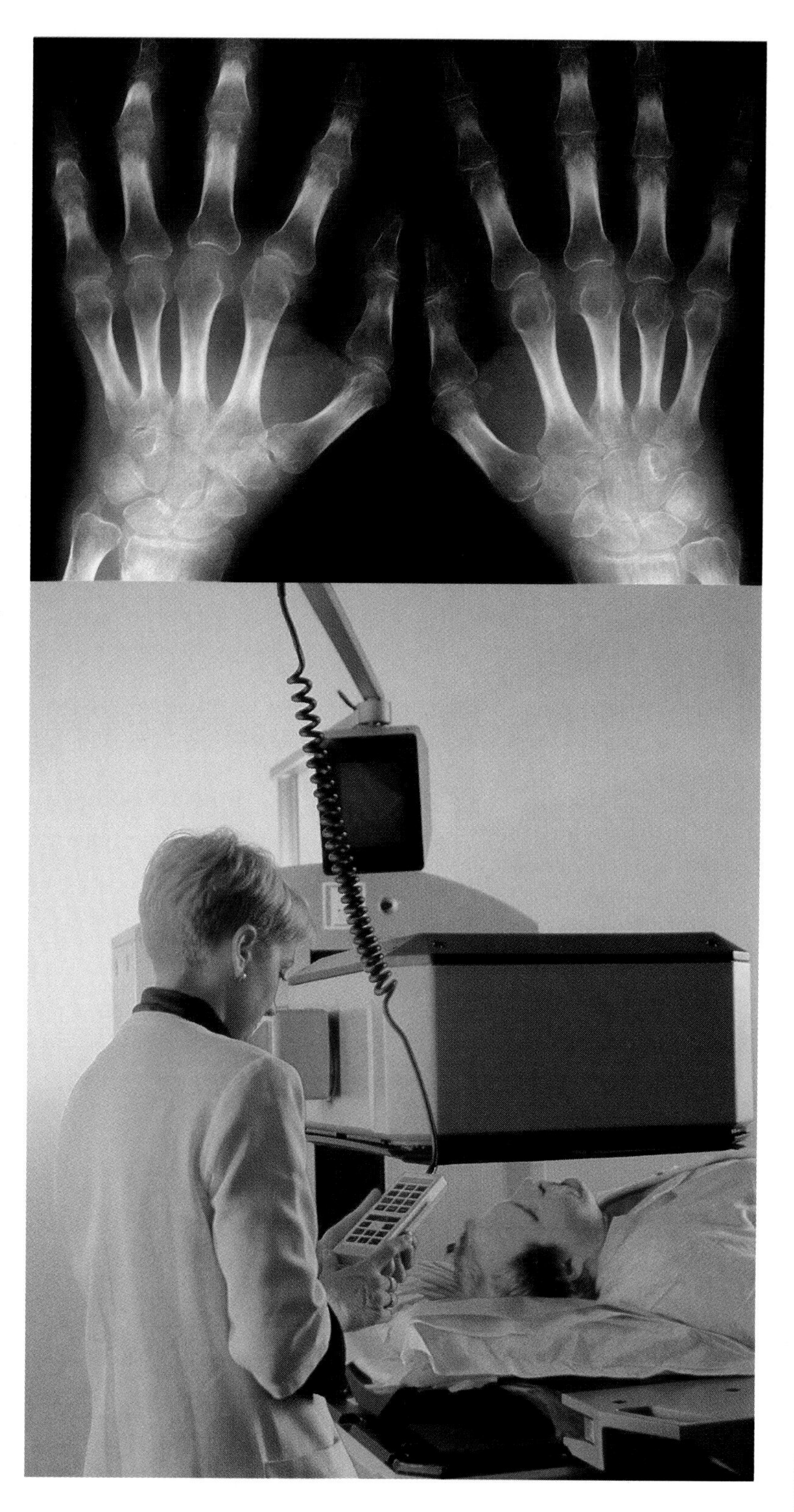

Top: An X-ray image of human hands. Useful for showing bones, X-rays are not helpful for other injuries or diseases, and can be injurious if improperly used. Bottom: The use of gamma scans helps detect and manage cancerous growths in the body. This machine takes pictures of a radioactive "tracer" as it concentrates in affected areas.

Particle Accelerators

How can one trap something as ephemeral as a subatomic particle and pin it down long enough to study how it behaves? Much of what we know about the fundamental constituents of matter comes through the use of particle accelerators. Known as "atom smashers" in popular parlance, these instruments use electric fields to accelerate particles and slam them into targets—all for the purpose of studying how they behave. Besides their obvious research potential, particle accelerators produce isotopes for medical, commercial, or military use.

Scientists may begin a typical accelerator experiment by firing a laser that excites particles in a source material, sparking subatomic particles—such as electrons or protons—to break off the material and shoot down a chamber, where their travel is accelerated by electromagnetic waves. If the chamber is linear, the particle simply shoots straight down to the end and smashes into a waiting target, which may be no more exciting than a sheet of foil. If the chamber is circular, a particle can spin indefinitely, gaining speed like a race car going around a racetrack, until scientists introduce a target for the particle to slam into. In the most powerful accelerators, two beams of particles collide with one another head-on to produce even greater collision energies. In the spray of debris from these collisions, physicists may study the behavior of matter's tiniest components or create entirely new elements or isotopes. The collisions are recorded with sensitive light detectors for analysis.

Particle accelerators are found in laboratories all over the world. They have aided the discovery of such particles as quarks, leptons, hadrons, and bosons. Some accelerators study only one particle, such as protons, and accelerators all differ in design, reflecting varied engineering approaches or research intents. Accelerators can be small enough to fit on a laboratory bench top, or as large as a small city. The world's largest—the Large Hadron Collider at CERN Laboratory in Geneva, Switzerland—is an underground circular accelerator with a tunnel-like chamber that is nearly 17 miles in circumference!

The first devices used to study atomic particles were accelerators by design, but were

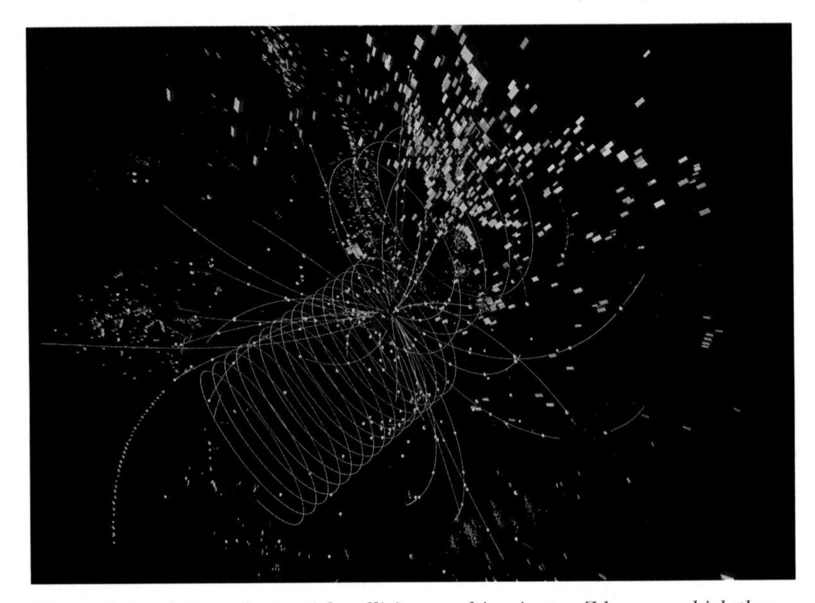

Above: A simulation of a particle collision resulting in two Z bosons, which then decay. Simulations help scientists predict results, measure successes, and provide insight into the minutiae of an event. Top left: Television uses a simple form of particle accelerator to produce its images.

not called by that name at the time. The first circular accelerator, however, is indisputably credited to Ernest O. Lawrence (1901–58), an American physicist, who sketched the design for his "cyclotron" in 1929. He built a simple, four-inch device that paired two D-shaped magnets in such a way that when he applied 2,000 volts of electricity to the apparatus, the particles in the chamber were induced to rotate in spiral fashion at kinetic energies of 80,000 electron volts, until they collided with a target on the outer edge of the chamber.

Large modern accelerators are constructed underground to conserve costs and to better contain the radiation released during experiments. The circular accelerator at Fermilab in Batavia, Illinois, is constructed below a 6,800-acre swath of land. Billed as the "world's highest-energy particle physics lab," Fermilab's accelerator whizzes particles around the facility's four-mile-round tunnel 50,000 times in one second, reaching energies more than 10 million times greater than those of Lawrence's first machine. Standing aboveground, you would hardly suspect that cutting-edge science is taking place below your feet, or that scientists have discovered three major new particles—top quarks, bottom quarks, and tau neutrinos—on this site. The laboratory's work continues, and the prairie land above is currently being restored to the state in which it existed before Europeans settled the region, with native plants and bison herds.

Scientific teams sign waiting lists, sometimes years in advance, for the chance to conduct an experiment at a specific, modern-day accelerator. They spend their time viewing tests in a closed room filled with computer monitors. But one need not be a scientist to operate a particle accelerator. There is one in the living room: The television set! When electrons are fired at phosphor impregnated on the back of a TV screen, the resulting reaction is a tiny dot of colored light. Our eyes read all these "pixels" as part of a moving picture.

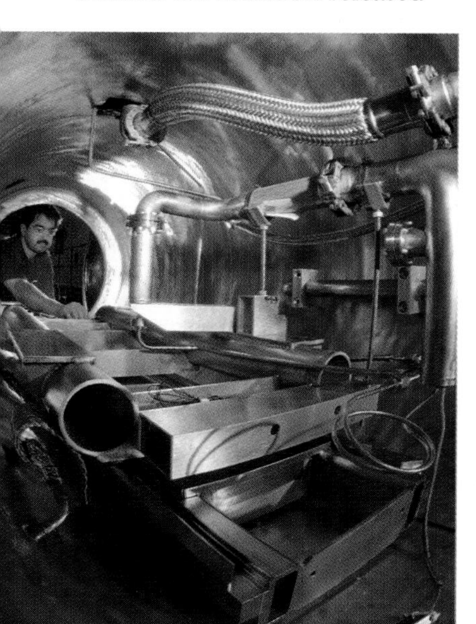

MP9 room, Fermi National Accelerator Laboratory (Fermilab), Batavia, Illinois. *The International Linear Collider (ILC) cryostat pictured above is currently being developed and will one day collide electrons and positrons at immensely high speeds.*

d mium .24	61 Pm Promethium (145)	62 Sm Samarium 150.36	63 Eu Europium 151.964	64 Gd Gadolinium 157.25	65 Tb Terbium 158.9253	66 Dy Dysprosium 162.5	67 H Holm 164.9
J ium 0289	93 Np Neptunium (237)	94 Pu Plutonium (244)	95 Am Americium (243)	96 Cm Curium (247)	97 Bk Berkelium (247)	98 Cf Californium (251)	99 E Einstei (25

The periodic table is a chart of all the known elements, their atomic numbers, symbols, names, and weights. It is frequently updated as new elements are discovered.

WELCOME TO THE CLUB

A rite of passage in high school is memorizing or at least familiarizing oneself with the periodic table of elements. But every year, this challenge gets tougher! Why? With a little help from particle accelerators, scientists are almost routinely discovering…no, *creating* new elements. In 2006, a team of Russian and American scientists confirmed that they that had detected a new, superheavy element—the heaviest ever—coming in at number 118 on the periodic table. Temporarily named *ununoctium,* it is a fleeting, temperamental element that doesn't allow itself to get too attached to anything. In fact, it self-destructs in less than a millisecond. Pay close attention: In the time it takes to read this book, the periodic table may have changed yet again.

Four Forces

Natural events, including disasters, are often casually referred to as acts of nature or an example of nature's power. An earthquake, hurricane, or any other major—and often destructive—natural event reminds us that there are great forces, often unpredictable and certainly uncontrollable, that exert great power over our lives. But nature's real power is not event-specific. It is being exerted all the time on both the tiniest and most expansive of scales, and it is not merely destructive but is actually holding the entire universe together.

AND THEN THERE WERE FOUR

There are generally considered to be four main forces in nature without which the basis for all matter, for all beings—the atom—would not exist. These four forces are responsible not only for governing these invisible-to-the-eye interactions but also the relationship between planets, stars, and suns we've yet to see. The idea of a "force" is in itself something that needs to be pondered before the various kinds of forces can be fully appreciated. To "force" someone to do something is often described as putting pressure on them, exerting some sort of influence over them and a particular decision they need to make or action they need to do. The four forces do not act on all matter. Gravity acts on all particles, electromagnetism only on charged particles, the strong force only on quarks, and the weak force on all matter particles. Influence is an effective way to think of the forces, as different forces will exert different influences. The force, the influence, determines how the matter will behave. Each force is associated with a specific particle, a kind of ambassador of that force that carries it over the range in which it can exert its influence. These particles are collectively referred to as gauge bosons.

ELECTROMAGNETIC FORCE

The idea behind electromagnetic force is most easily described—and has been in prior chapters—as "opposites attract." The converse is also true, that like repels like. Positive ions are happily attracted to negative ions in a solution, but are not so eager to join up with other positive ions. The electromagnetic force is carried between particles by photons, which also make up light rays.

STRONG NUCLEAR FORCE

Strong nuclear force is considered to be the strongest of these four formidable forces and is carried by particles called gluons. The strong nuclear force is essentially what holds the nucleus together and counteracts the electromagnetic forces in stable nuclei, which on their own would break the nucleus apart, since the positive charges of all the protons would repel each other.

WEAK NUCLEAR FORCE

Weak nuclear forces, carried by W and Z or weak gauge bosons, explain the beta decay of radioactive isoptopes such as carbon 14, uranium 235, and potassium 40. For example, as uranium 235 undergoes radioactive decay and turns into lead 207, the neutron of the uranium radioisotope decays into a proton.

GRAVITATIONAL FORCE

When Isaac Newton got his famous bump on the head while sitting under an apple tree, he merely applied acquired knowledge to a phenomenon that he and everyone before him had seen for thousands of years. Though gravity is the weakest of the forces, it is the one that is the most tangible: the arc of a Frisbee after it's

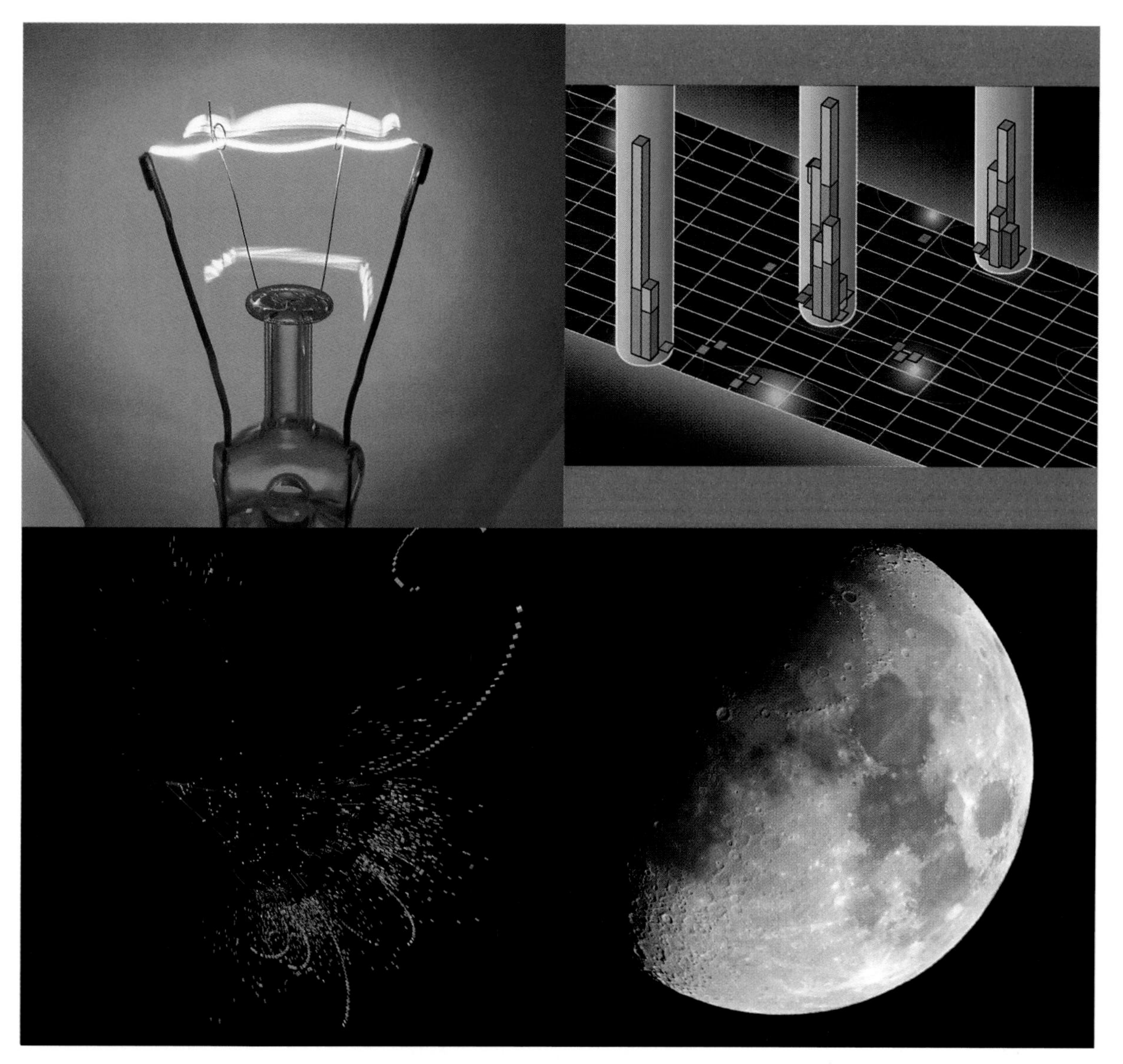

Top left: A glowing lightbulb is an everyday demonstration of the electromagnetic force. Top right: A depiction of the quarks and gluons that make up protons, as detected in the Collider Detector at Fermilab. Bottom left: A simulation of the production of a Higgs particle and a Z boson, which carries the weak nuclear force. Bottom right: The influence of the Moon's gravity can be seen on Earth in the regular rise and fall of the ocean's tides. Opposite page, top left: Every action we undertake, even that of throwing a Frisbee, is affected by the forces that govern the universe.

thrown, rises, and then falls; a glass teetering too close to the edge of a table that topples to the floor. Counteracting the comparative weakness of gravity in respect to the other forces are two factors: its additive property and its range. Adding any weight whatsoever to an object affects its gravitational force, and gravity is the one force that is effective over any range—it is virtually limitless. Gravity does not just exert its force on the relationship between the Frisbee and the grass, but also on the relationship between the Sun and Earth. It has been proposed that gravity is carried by gravitons, but their existence has yet to be confirmed.

Formation of the Elements

In the recesses of space, in the dark expanse of the endless universe, are formed the elements that compose our entire existence. Much like the way molten steel takes on a completely new shape and properties within the fiery innards of a furnace, stars are a forge in which the new elements are created, the elements that make up the air we breathe, the matter that surrounds us, and our very own beings. The elements that make up our bodies—the carbon in our organic chains, the oxygen carried by the red blood cells through our arteries—came from stars.

IT STARTED OFF WITH A BANG

According to the big bang theory, there was a lot going on in the moments immediately following the blast. Atomic particles floated about in the cosmos, stewing at around a sweltering 10 billion degrees. Hydrogen, helium, and some lithium are said to be the first elements that were formed on the scene, and hydrogen and helium remain the most abundant elements in the universe. Hydrogen alone is believed to account for the majority of the material found from here to the far reaches of space.

This still leaves a lot of elements unaccounted for. Other than hydrogen and helium, it is widely accepted that most elements are created in stars through a process called nucleosynthesis. Nucleosynthesis is just what it sounds like—the synthesis of a nucleus. The concept of stellar nucleosynthesis got a big boost from the work of Nobel Prize-winning physicist Hans Bethe. Bethe's work—which was famously published in 1939 in a paper titled, "Energy Production in Stars"—focused primarily on the thermonuclear reactions that fuse hydrogen together to form helium. In the decades following Bethe's work, his theories were further expanded upon by many scientists, including Margaret and Geoffrey Burbridge, William Fowler, and Fred Hoyle, who, in 1957, published "Synthesis of the Elements in Stars." All of the research in this area, which began before Bethe's paper, is constantly being reevaluated and explored. With all that scientists know about the universe, so much of what composes it—dark matter and dark energy, for example—is still unknown.

HOT STARS

The Sun, which is itself a star, performs this heated fusion or nucleosynthesis constantly. Deep in the core of the Sun, hydrogen atoms are fused into helium, and in the process a mind-boggling amount of energy is released. And even though energy is released during the process of fusing hydrogen, the fusion has to take place at a minimum temperature of roughly five million degrees. This reaction takes place in all other stars as well, as long as there is hydrogen available, of course. After stars have exhausted their supply of hydrogen, they will actually begin fusing or "burning" the helium. These processes can form heavier elements like nitrogen or oxygen and are still highly exothermic, releasing large amounts of energy, but only up to a point. Beyond the element iron, the process of fusing together nuclei

Above: William C. Fowler, a twentieth-century physicist whose work contributed to understanding stars and the role they play in the creation of the elements found in the universe. Top left: Galaxies like this distant one are composed of billions of stars.

The Crab Nebula, as viewed with the Hubble Space Telescope, is the remnants of an exploded star and as such is one place where some of the heaviest elements found in nature might originate.

to form the heavier elements becomes endothermic—requiring more energy to be put into the fusion reaction than will be given off. Stars much larger than the Sun—which is a relatively small star in comparison to other heavyweights in the universe—are needed to form elements heavier than oxygen. But stars can't go on forever. It is believed that when stars explode, they do so because they have essentially gone through all their fuel, having already made as much carbon, or nitrogen or nickel, as they can make. They are quite literally burnt out. These exploding stars—that pack a blast of around 10^{28} megatons—are referred to as supernovas, and are believed to result in the formation of a wide variety of the heavier elements in the universe.

Famous Chemists

Nikola Tesla (1856–1943)
Serbian inventor and electrical engineer. Tesla experimented with electricity and magnetism, inventing a method of generating electricity by alternating current (AC). Though initial acceptance was rocky, AC power revolutionized the world and is the chief method of power generation used today.

Michael Faraday (1791–1867)
English chemist and early investigator of magnetism. Faraday is regarded as one of the greatest experimental scientists ever. Faraday discovered electromagnetic rotation—the fact that electricity is generated when magnets are rotated in proximity to each other. He also outlined the induction theory of heat, conceived the Second Law of Electrolysis, and invented the dynamo, among many other scientific tools.

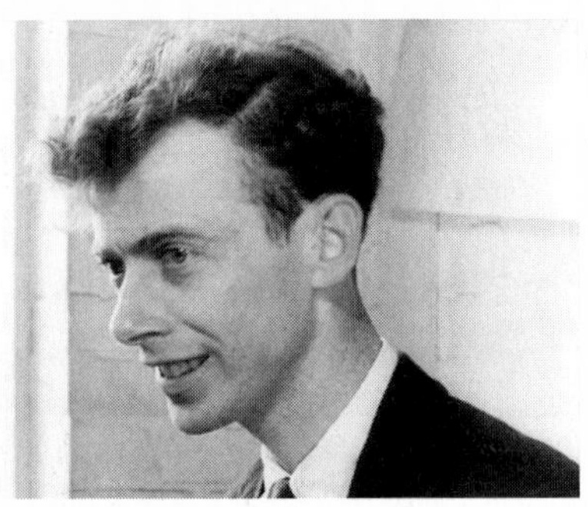

James Watson is known best for his work with DNA structure.

James Watson (b. 1928)
and Francis Crick (1916–2004)
American and English DNA scientists. Watson and Crick are credited with deducing the double helix structure of DNA and publishing that finding in 1953. The pair produced a three-dimensional model, which showed scientists for the first time how genetic material is replicated and passed on to succeeding generations.

Rosalind Franklin (1920–58)
English chemist. Franklin's X-ray diffraction photographs of DNA greatly influenced the work of Watson and Crick. Franklin never received the Nobel Prize for her DNA work because she died before the prize was awarded in 1962.

Marie Curie, winner of two Nobel Prizes.

Marie (1867–1964) **and Pierre Curie** (1859–1906)
Husband-and-wife team of scientists, based in France, who were pioneers in the field of radioactivity. The Curies discovered and named the elements radium and polonium. Polish-born Marie was the first person to win two Nobel Prizes; the first to win them in two separate fields, chemistry and physics; and the only woman to have won two Nobels.

Ernest Rutherford (1871–1937)
New Zealand-born nuclear physicist. Hailed as the father of nuclear physics, Rutherford was the first to advance the orbital theory of the atom, namely, that atoms consist of nuclei orbited by a cloud of electrons. He named alpha, beta, and gamma rays.

Linus Pauling (1901–94)
American chemist. Regarded as the top chemist of the twentieth century, he applied principles of quantum mechanics to the world of chemistry. Pauling won two Nobel Prizes—one for his work on the nature of chemical bonds, and another for his work promoting peace.

Twentieth-century American chemist Linus Pauling.

Joseph Priestley (1733–1804)
English chemist. Priestley is credited with the discovery of oxygen. Though he was not the first to discover the element, he was the first to publish his finding. Neither he nor Carl Wilhelm Scheele, oxygen's other discoverer, realized that their finding was a chemical element. Priestley called his discovery "de-phlogisticated air."

Alfred Nobel (1833–96)
Swedish chemist and inventor of dynamite. Though he did not invent nitroglycerin, Nobel did perfect an efficient method for commercial production of the explosive. Wishing to correct his legacy as a "merchant of death," Nobel left his fortune to establish the Nobel Prize, the world's highest honor for achievement in literature, the sciences, economics, and the cause of peace.

Antoine-Laurent Lavoisier laid the foundation for the work of modern chemists.

Antoine-Laurent Lavoisier (1743–94)
French chemist. Established that matter is neither gained nor lost during a chemical reaction. Demonstrated the role of oxygen in combustion and reduction reactions. Named "oxygen," taking its name from the Greek for "acid-former." A businessman and financier during the day, he is widely considered to be the father of modern chemistry. He was beheaded during the Reign of Terror of the French Revolution.

Robert Boyle (1627–91)
Irish chemist, natural theologist, and philosopher. Studied and wrote about the nature of air. Established what would come to be known as Boyle's Law, which states that any given volume of gas kept at a constant temperature will vary inversely in relation to its pressure.

Amedeo Avogadro (1776–1856)
Italian chemist. Stated that equal volumes of different gases kept at the same temperature and pressure contain the same number of molecules. This number is known as Avogadro's number, and is 6.022 x 10^{23}. This constant is used to determine the relative molecular weights of gases.

Amedeo Avogadro, known for his theories on molecular weight.

Dmitri Mendeleev (1834–1907)
Russian chemist. Studied the properties of the elements and devised a system for categorizing them according to their similarities and differences. This led him to create the first periodic table of the elements, which included blank spaces for elements that had yet to be discovered.

Niels Bohr (1885–1962)
Danish physicist. Noted contributor to the field of atomic physics. Developed a model of the atom that depicted the particle with a small, positively charged nucleus, surrounded by negatively charged electrons that traveled in discrete orbits of quantized energy and held together by electrostatic forces.

English physicist and chemist John Dalton.

John Dalton (1766–1844)
English chemist. Dalton's law is named for him, and states that the total pressure of a homogenous mixture of gases is the same as the sum of the partial pressures of the individual gases present. Performed research in the field of colorblindness and also suffered from it. Colorblindness is sometimes referred to as "Daltonism."

Rachel Carson (1907–64)
Marine biologist, zoologist, and author. Wrote *Silent Spring*, a landmark book that examined the role of chemicals in agricultural science and sparked the modern environmental movement. Testified before Congress about the importance of protecting both human health and the environment, and that the two were inextricably linked.

Rachel Carson, author of Silent Spring.

Robert Oppenheimer (1904–67)
American physicist. Often referred to as the "father of the atomic bomb," Oppenheimer was director of the Manhattan Project, which was in operation during World War II and developed the first nuclear weapons. He served as chairman of the U.S. Atomic Energy Commission and opposed the development of the hydrogen bomb. Accused of being a communist sympathizer during McCarthyism.

Milestones in Chemistry

9000–8000 bce
Evidence of copper first being used in Middle East. Copper occupies the stage between the Neolithic Age and the Bronze Age and was believed to be used in tools and weapons. Copper tools are believed to have been used to construct the great pyramids of Egypt.

8000 bce

A sixteenth century woodcut depicting brewers at work.

The process of fermentation is used to make beer and wine. 5000–6000 BCE: Copper smelting is believed to have first occurred in the Middle East.

4200–3200 bce
First copper alloys made to make bronze. Initially copper and arsenic were used, followed by the more efficient and more durable copper and tin mix, which is still used today.

2500 bce
Egyptians use sand, limestone. and soda to make glass.

c. 1200–1000 bce
Transition to the Iron Age as iron begins to replace bronze on a large scale.

1000 bce
Enzymes are used to make cheese and milk.

c. 460–c. 370 bce
Life of Democritus, ancient Greek who put forth the idea that all matter was composed of tiny discrete particles that could not be seen. He called these particles "atomos."

c. 384–322 bce
Life of noted Greek philosopher Aristotle, who supported the theory put forth by Empedocles that everything was made of earth, air, fire, and water. It is believed that in *De Philosophia,* Aristotle added "aether" to the mix, which he proposed made up the heavens. The four- and five-element theory was much more popular than the Democritus atom theory and greatly influenced science and the development of alchemy.

Late 300s ce
Roman Emperor Diocletian bans alchemist writings in the Roman Empire.

c. 721–c. 815
Life of Jabir ibn Hayyan, also known as Geber, influential Islamic alchemist.

980–1037
Abu Ali ibn Sina of Iran, called Avicenna, refutes idea that base metals can be made into noble metals including gold. Avicenna is believed to have influenced the ideas of Western alchemists.

Avicenna

1000
Gunpowder is invented in China.

Smokeless or "gun" powder.

1206–1280
Albertus Magnus, or Albert the Great. Patron saint of natural sciences, he brought Greek and Arabic science and philosophy to Europe. His life and work demonstrated that science was a fit subject of study for Christian scholars.

1493–1541

Life of Theophrastus Bombastus von Hohenheim, who is more commonly known as Paracelsus. Paracelsus believed in the use of alchemy to treat the ill.

1648

The collected works of Belgian scientist Jan Baptista van Helmont (1580–1644) are published. Van Helmont is credited with coining the word "gas," which he considered a substance distinct from solids and liquids.

1661

Robert Boyle published *The Sceptical Chymist*, in which he essentially redefines an "element" as a substance that cannot be broken down into any other substance.

1662

Boyle's law is published. Boyle discovered that the product of the volume and pressure of a fixed amount of gas is constant at a fixed temperature.

Late 1600s: Johann Becher and Georg Stahl develop the phlogiston theory, which attempts to explain the release of gases during combustion. The theory was overturned by Lavoisier but galvanized science for 100 years.

1714

Gabriel Fahrenheit develops the first mercury thermometer. He also develops the Fahrenheit scale of temperature.

1756

Scottish chemist Joseph Black discovers carbon dioxide, which he calls "fixed air."

1752

Ben Franklin flies his famous kite and discovers a "spark," which proves his idea that lightning contains electricity.

1766

Henry Cavendish discovers hydrogen.

Benjamin Franklin flying his famed kite.

1772 and 1774

Discovery of oxygen. Swedish scientist Carl Wilhelm Scheele discovers oxygen in 1772 but does not publish his results until 1777. In 1774, Joseph Priestly independently discovers oxygen and publishes his works before Scheele.

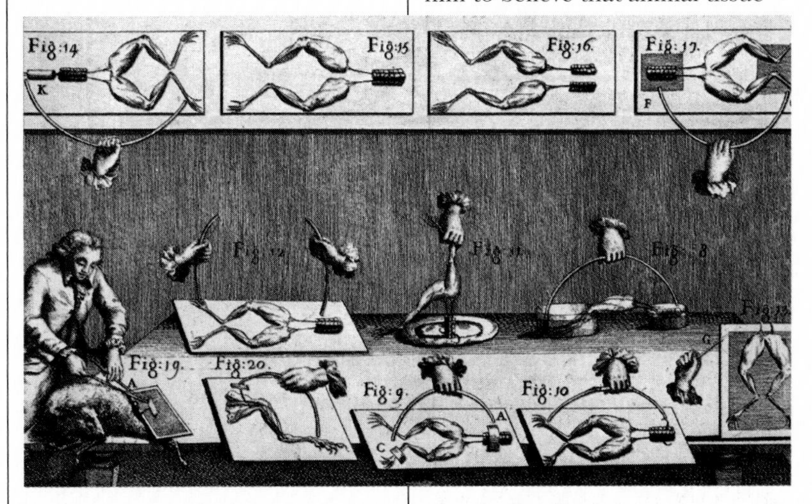

Galvani experimenting with electric currents.

1787

Antoine-Laurent Lavoisier, Claude Louis Berthollet, Antoine de Fourcroy, and Guyton de Morveau publish *Méthode de nomenclature chimique* (method of chemical nomenclature).

1791

Luigi Galvani notices that the muscles in the severed legs of a frog twitch while resting on a metal plate. His experiments lead him to believe that animal tissue generates electricity. His thesis was originally accepted by colleagues but not accepted by Alessandro Volta, who was inspired in his own investigations into electrochemical cells.

1794

Antoine-Laurent Lavoisier is beheaded at the guillotine during the French Revolution. He is credited by many as being the father of modern chemistry. He was known for his study of the elements and identified oxygen as one of them, giving it its name and explaining its role in combustion and plant and animal respiration. He also classified oxygen as an acidifier. His book *Traité élémentaire de chimie* (treatise on elementary chemistry), published in 1789, elaborated upon his theories on the properties of elements.

1796

Edward Jenner develops the first vaccine, used against smallpox.

1800

Alessandro Volta shows that two different metals separated by a moist conductor will conduct a charge. His stacking of zinc and silver plates—the voltaic pile—to generate an electric charge was the precursor to modern batteries.

1803

John Dalton first presents his findings on differentiating between the atoms of different elements by their atomic weights, and later publishes his theories in 1808. His work, which describes the attributes of atoms as unique to their elements, is the basis for modern atomic theory.

1811

Amedeo Avogadro hypothesizes that if equal volumes of gases are kept at the same pressure and temperature, they will contain the same number of molecules. This theory is later called Avogadro's law.

1820

Hans Christian Ørsted demonstrates that an electric current produces a magnetic field.

1831

Michael Faraday begin his experiments that eventually lead to the discovery of electromagnetic induction using magnets to induce an electric current. The Law of Induction is attributed to Faraday, and the unit of electric charge, the faraday, is named for him.

1846

First surgical operation using ether as an anesthetic is performed at Massachusetts General Hospital in Boston.

1848

Sir William Thomson, also known as Lord Kelvin of Scotland, devises the absolute temperature scale, placing zero degrees as the lowest possible temperature at which point all atomic motion ceases. 1856: The first synthetic dye, mauve, is developed by William Perkin, revolutionizing the dye industry.

1859

Gustav Robert Kirchhoff and Robert Wilhelm Bunsen observe that different substances emit different spectra when they are burned, a significant advancement in the burgeoning field of spectroscopy.

1867

Alfred Nobel patents his nitroglycerine-based explosive material dynamite after years of experimentation. The wealth he earned in his lifetime he left upon his death to the establishment and administration of the Nobel Prize.

1869

Dmitri Mendeleev publishes the first periodic table of the elements.

A lightbulb.

1879

Thomas Edison perfects the incandescent lightbulb, building on more than 50 years of prior research by other scientists.

1888

Inventor Nikola Tesla presents his alternating current (AC) system of generators, which catch the attention of George Westinghouse; their partnership helps bring electricity to homes across America.

1895

Wilhelm Röntgen discovers X-rays.

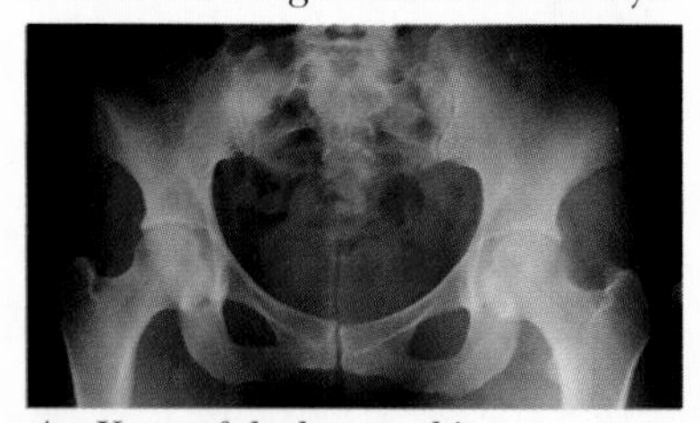

An X-ray of the human hip.

1896

Antoine-Henri Becquerel discovers and publishes a paper on spontaneous radioactivity, though the word "radioactivity" is later coined by Marie Curie.

1897

Chemist Felix Hoffman develops a chemically pure form of acetylsalicylic acid. His employer, Friedrich Bayer, markets the product as aspirin.

1897

Joseph John Thomson discovers the subatomic particle known as the electron.

1903

Becquerel and Pierre and Marie Curie share the Nobel Prize for their work in radiation.

1909

Danish scientist Soren Sorensen develops the pH scale.

1911

Ernest Rutherford publishes his theory of the structure of the atom as having a nucleus with a positive charge that is surrounded by negative electrons.

1913

Niels Bohr expands on Rutherford's atomic theory with his own model of the atom in which electrons circle the nucleus in distinct orbits.

1913

Frederick Soddy develops his concept of isotopes, describing them as different forms of the same element with different atomic weights.

1920

Rutherford gives the name "proton" to the hydrogen nucleus, and predicts the existence of uncharged nuclear particles, which would later be called neutrons.

1928

Alexander Fleming, while working on an influenza virus, comes across a mold that seems to kill bacteria. He names the substance penicillin.

1932
James Chadwick, onetime pupil of Ernest Rutherford, discovers the neutron.

1935
Wallace Carothers develops nylon, a synthetic polymer, which revolutionizes the textile industry.

1939
Paul Müller synthesizes the insecticide dichlorodiphenyltrichloroethane, more commonly known as DDT.

1939–1945
The drive to develop and produce

A tire made from synthetic rubber.

synthetic rubber is sparked by the advent of World War II, during which time the United States is cut off from its primary supply of natural rubber in Southeast Asia. During this period, the mass production of synthetic rubber blossoms, and the first car tires using synthetic rubber hit the road.

1945
First atomic—nuclear fission—bomb is built and detonated in New Mexico. Robert Oppenheimer heads the Manhattan Project, which oversaw the creation of the weapon. Later that same year, atomic bombs are dropped on both Hiroshima and Nagasaki.

1952
Jonas Salk produces the first vaccine to combat polio.

1953
Francis Crick and James Watson build and present their double-helix model of DNA, which is based on the work of Maurice Wilkins and Rosalind Franklin.

1961
Jack Kilby and Robert Noyce develop the first silicon chips.

1963
Rachel Carson publishes her controversial book *Silent Spring*, which eventually results in the banning of the insecticide DDT. Her book inspires the modern environmental movement.

1964
Using a radio telescope, Arno Penzias and Robert Wilson detect cosmic background radiation, the leftover radiation remnants from the big bang.

1976
The commercial availability of the catalytic converter changes automobile emissions forever, as the converter partially "cleans" car exhaust. The converter was originally developed by John Mooney and Carl Keith.

A catalytic converter.

1977
Magnetic resonance imaging (MRI) is used for the first time on a human patient.

1985
Buckminsterfullerenes are discovered by Richard Smalley, Harold Kroto, and Robert Curl. These "buckyballs" are named for architect Buckminster Fuller's geodesic dome, and contain 60 carbon atoms bound together in the shape reminiscent of a soccer ball.

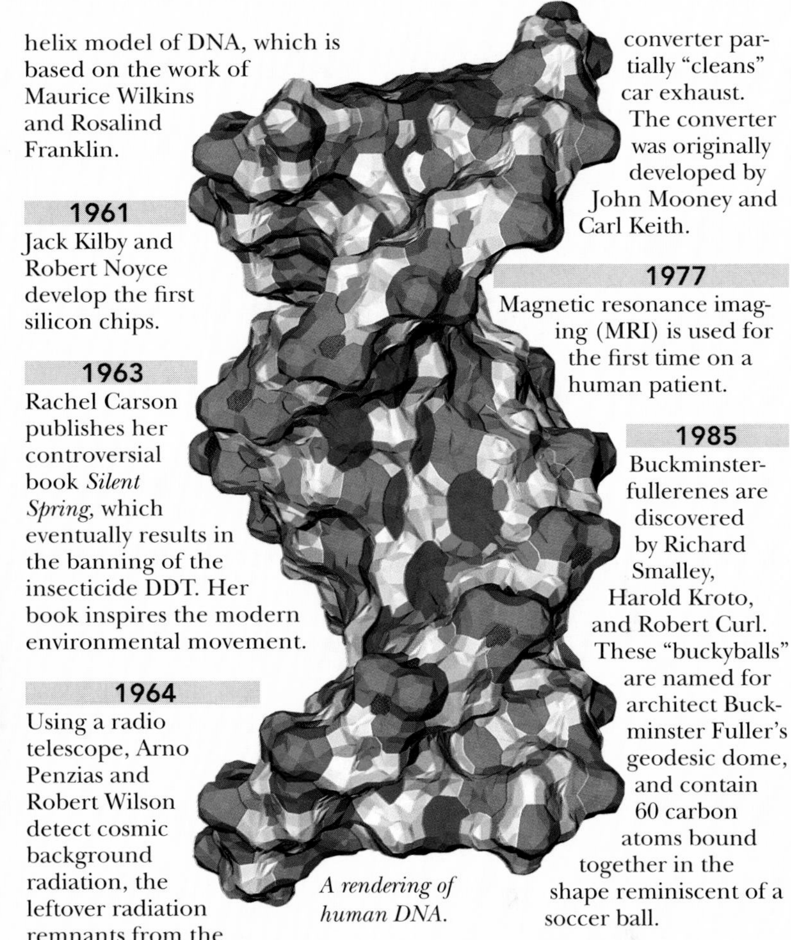

A rendering of human DNA.

mid-1980s
The Human Genome Project begins with the purpose of sequencing and mapping the human genome.

1996
Dolly the sheep becomes the first mammal to be cloned successfully. She lives for six years. Many large mammals have since been cloned.

2003
Scientists complete the full sequencing of the human genome and publish their findings.

The Periodic Table

Russian chemist Dimitri Mendeleev is credited with creating the first periodic table of the elements. Using the results of his dogged investigation into the varying properties of the individual elements, he organized them into table form based on their similarities and differences. Mendeleev's goal was to arrange the elements according to their atomic weights, based on the assumption that the changes in properties would follow the changes in atomic weights. He referred to this as the "periodicity of the elements."

The Modern Periodic Table

The table is arranged in columns and rows, with the atomic numbers increasing from left to right and from top to bottom. Each row represents a different "period," while columns comprise the different groups.

All the elements in a specific period have the same number of atomic shells, also called orbitals. Atomic shells represent the positioning of electrons at fixed distances from an atom's nucleus and indicate the amount of energy electrons possess at that distance—electrons closer to the nucleus have lower energy than electrons that are located farther away (see chapters one and two).

In the second row or period of the table, all the elements have two shells. That means that both lithium and carbon have the same number of atomic shells, although in their neutral states they have different numbers of electrons (three and six, respectively). Moving from left to right in a given row also represents a shift from metals to nonmetals.

All the elements in a given column form a group. Members of the same group all have the same number of electrons in their outer shell. For example: Beryllium and barium are both members of group IIA, and both have two electrons in their outermost shell. There are some exceptions to this trend in

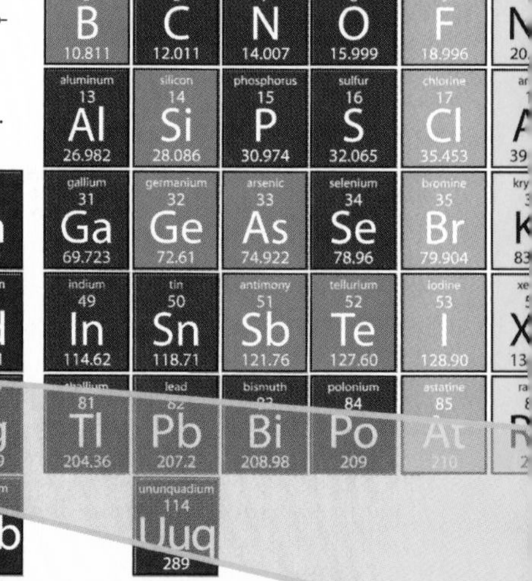

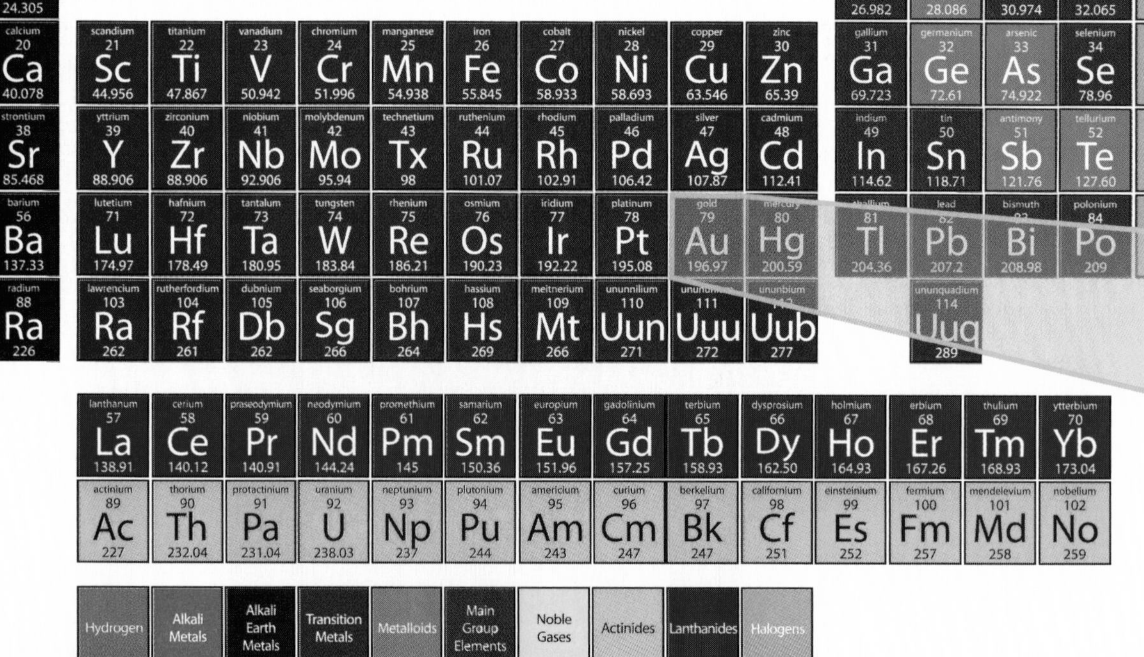

The periodic table. The element gold (Au) is highlighted for a closer look.

"I began to look about and write down the elements with their atomic weights and typical properties, analogous elements and like atomic weights on separate cards, and this soon convinced me that the properties of elements are in periodic dependence upon their atomic weights," Mendeleev wrote in *Principles of Chemistry.*

Glenn Seaborg, the American physicist who won the 1951 Nobel Prize for Chemistry.

the transition metals found at the center of the table. The transition metals have a structure that makes them capable of placing valence electrons—electrons in the outermost shell—in more than one shell. The chemical properties of an element are largely determined by the structure of its outer shell, so elements in the same group will have similar chemical behavior, just as Mendeleev noticed.

Families

The main families of the periodic table are the alkali metals, alkaline earth metals, transition metals, rare earth metals, halogens and noble gases, as well as a few unnamed families containing both metals and nonmetals. The elements of a given family share similar properties, both physical and chemical. For example: Running down the far right-hand side of the periodic table are the inert or noble gases. They have completed outer shells, making them less likely to react with other elements (inert).

Below the main portion of the periodic table are two rows. The top row is the lanthanide series, while the bottom row is the actinide series. Elements in the lanthanide series are also known as the rare earth metals and occur naturally, but only in very small amounts. Many of the elements in the actinide series are used in nuclear reactions, and many of them do not occur naturally, but are instead created in a laboratory setting.

About the Elements

Each element is represented by a symbol. The names and symbols of the elements are based on a variety of factors. Lead, for example, is represented by the symbol "Pb," derived from its Latin name, *plumbum.* (This is also the origin of the word "plumber," as pipes were made of lead.) Polonium, on the other hand, was discovered by Marie Curie and named for her home country, Poland. Other elements are named for people: Seaborgium (Sg) was named for physicist Glenn Seaborg, who himself helped discover 10 elements and proposed adding the actinide series to the periodic table, making him instrumental in developing its current arrangement.

Along with the symbol, each box contains the element's atomic number in the top left- or right-hand corner. This is the number of protons in an atom of that element. The number below the symbol is the atomic mass, or the number of protons and neutrons in an atom. Since some elements can exist naturally in various isotopic forms—varying numbers of neutrons—this mass represents an average of the various isotopes of the element and their frequency. For example: The most common isotope of carbon has six protons and six neutrons giving an atomic weight of exactly 12—Carbon 12 is the basis for all atomic weights. However, there are two other naturally occurring isotopes of carbon: Carbon 13 has seven neutrons, and carbon 14 has eight neutrons. Although the vast majority of naturally occurring carbon has an atomic mass of 12, the atomic mass listed on the periodic table (12.010) reflects the existence of these other isotopes.

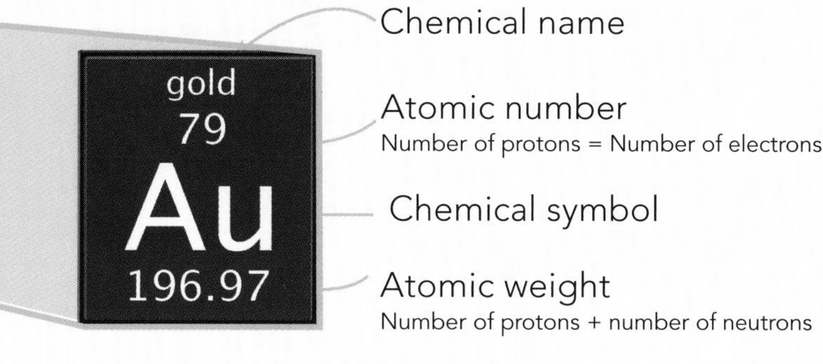

Inside the Atom

The word "atom" comes from the Greek word meaning indivisible. A popular image of the atom—and one that is helpful for a basic understanding of the atom, including its structure and how it works—is based on what was called the planetary model, which is exactly what it sounds like: the idea of smaller bodies orbiting around a larger, central one. John Dalton's atomic theory of atoms as indivisible units was elaborated upon by Ernest Rutherford and then Niels Bohr, whose popular Bohr model was eventually followed up by the quantum model of the atom.

The vast majority of the atom is actually empty space. The nucleus is at the center of the atom and consists of positively charged protons and neutral neutrons. Together they are referred to as nucleons, and they account for virtually all the mass present in the atom. Existing in what are commonly referred to as orbitals that circle the nucleus are the electrons, negatively charged particles.

Electron

Electrons are subatomic particles that have a negative electric charge. Most believe the electron to be a fundamental particle, meaning it cannot be broken down or split.

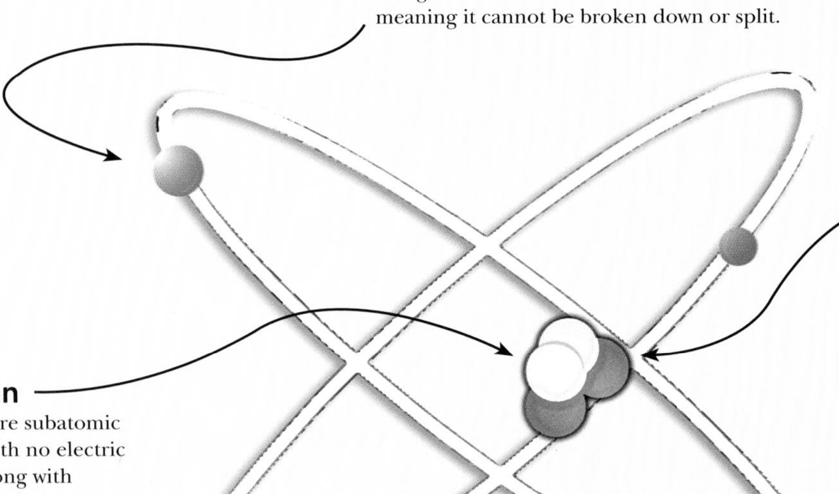

Proton

The subatomic particles with a positive electrical charge, protons reside in the nucleus of an atom. The helium (He) atom contains two protons.

Neutron

Neutrons are subatomic particles with no electric charge. Along with protons, neutrons reside in the nucleus of an atom. The isotope of an element is determined by the number of neutrons in its nucleus.

Orbital

The atomic orbital is the area surrounding a nucleus inhabited by an electron or electrons.

A diagram of the helium atom.

Because helium (He) is the second-lightest element—only hydrogen (H) is lighter— balloons filled with helium will rise in our comparatively heavier atmosphere.

In stable atoms, the number of protons is the same as the number of electrons, but atoms can lose, gain, or share electrons in various chemical reactions. An atom that loses an electron loses part of its electric charge and therefore becomes more positive. It is called a positive ion. An atom that gains an electron gains electric charge and therefore becomes more negative. It is called a negative ion. The number of neutrons in an atom can also vary, either naturally or by manipulation in a laboratory setting. Atoms of a particular element that have different numbers of neutrons are called isotopes. For example: Carbon 12 has six neutrons and six protons, while carbon 13 has seven neutrons and six protons.

Bonds

Atoms want to have "full" outer shells, shells that have no more room for additional electrons. This is when atoms are at their most stable. The noble gases all have eight electrons in their outer shells, which is what makes them so unreactive or "inert," as they are also called. If an atom is not "noble" by birth, so to speak, the way to achieve a completed outer shell is by bonding with another atom that also wishes to complete its outer shell. The electrons in the outermost shell of an atom are called valence electrons. These are the electrons that are involved in bonding. Electron bonds come in various forms:

Covalent: In a covalent bond, two atoms share electrons.

Polar covalent: In a polar covalent bond, the electrons are shared between two atoms, but the sharing is unequal and the electrons spend more of their time around one atom than the other.

Ionic: In an ionic bond, an atom that gives up the valence electrons bonds in its outer shell bonds with an atom that needs to gain electrons to complete its outer shell. Sodium (Na) and chlorine (Cl) do this to form salt.

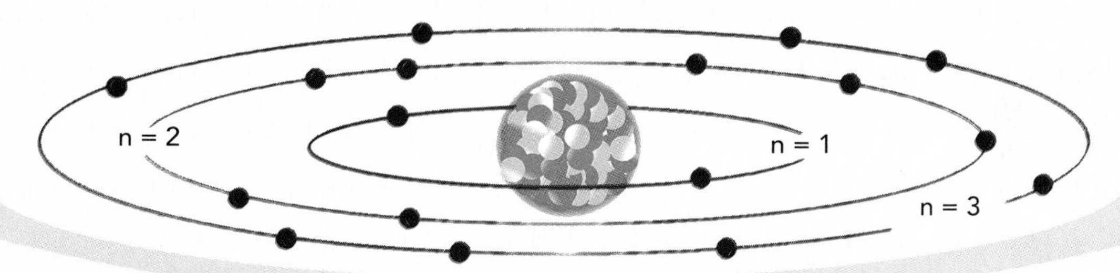

Finding Electrons

As physicists and chemists learned more about the structure of the atom, it was determined that the exact position of an electron in an atom could not be pinpointed. It is perhaps more accurate to describe electron orbitals as density clouds, and to think of orbitals as the place most likely to find an electron at any given time.

Electron orbitals can hold two electrons each. They come in different shapes and sizes and exist at different energy sublevels, generally determined by their distance from the nucleus. The different electron orbitals are sometimes referred to as sublevels, since together they make up a particular energy level.

The organization of an atom's electrons into levels and orbitals is called its electron configuration. The first number refers to the energy level; the letter refers to the type of orbital; and the last number refers to the number of electrons present. For example: A carbon atom contains six electrons. Its electron configuration is written 1s2, 2s2, 2p2. This means that at the first energy level, there are two electrons and they are in an "s" sublevel. At the second energy level, there are four electrons, two in an "s" sublevel and two in a "p" sublevel.

Important Numbers

1 gram (small) calorie	Amount of energy needed to raise 1 gram of water 1 degree Celsius. Equivalent to 4.184 joules.
1 large calorie	4.184 kilojoules
Angstrom	1×10^{-10} meters
Avogadro's number	6.022×10^{23}. The number of particles in one mole of a substance.
Becquerel	2.70×10^{-11} Ci (curies). Named for Henri Becquerel.
Bohr radius of an atom	$5.2917725 \times 10^{-11}$ m. Named for Niels Bohr.
Charge of an electron	1.6×10^{-19} coulombs
Curie	3.7×10^{10} decays per second in radiation. Named for Marie Curie.
Dalton (Da) or Unified Mass Unit	$1.66053886 \times 10^{-27}$ kg (1.66×10^{-24} g). Based on 1/12 of a carbon 12 atom. Named for John Dalton.
Faraday's Constant	96,485.31 C/mol. The total charge of one mole of electrons.
Freezing point of water	Degrees Celsius: 0 Degrees Fahrenheit: 32 Degrees Kelvin : 273.15
Joule	Energy required to exert a force of one newton for a distance of one meter. Named for James Prescott Joule.
Mass of a neutron	1.674929×10^{-27} kg
Mass of a proton	1.672623×10^{-27} kg
Mass of an electron	9.109390×10^{-31} kg
Nanometer	1×10^{-9}—or 1 billionth-meter
Planck's constant	6.626×10^{-34} J times seconds. Named for Max Planck.
Speed of light	186,282,397 miles per second or 670,616,629 miles per hour.
Watt	1 joule per second. Named for James Watt.

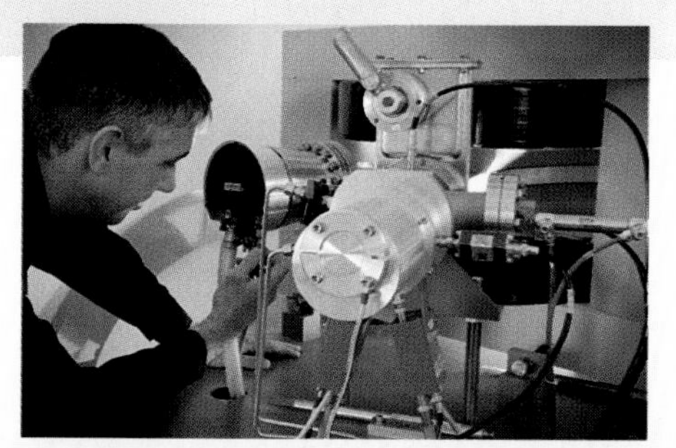

Right: The mass of subatomic particles can be measured using a mass spectrometer.

Left: When water temperature drops below the freezing point, ice is formed. Above: An artist's rendering of an atom.

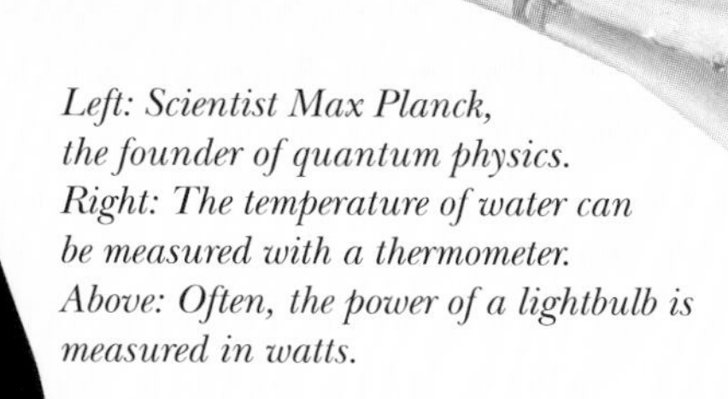

Left: Scientist Max Planck, the founder of quantum physics. Right: The temperature of water can be measured with a thermometer. Above: Often, the power of a lightbulb is measured in watts.

Living Through Chemistry

Elements in the universe recycle themselves constantly. They are absorbed, inhaled, or taken in by living organisms or nonliving processes, used for a short time, then released, to be used again. Earth is home to three important cycles. Without them, life as we know it would not be possible.

The Carbon Cycle

After hydrogen (H), helium (He), and oxygen (O), carbon (C) is the most abundant element in the universe. Tens of billions of tons of carbon are recycled every year by natural processes. Here's how it works in plants:

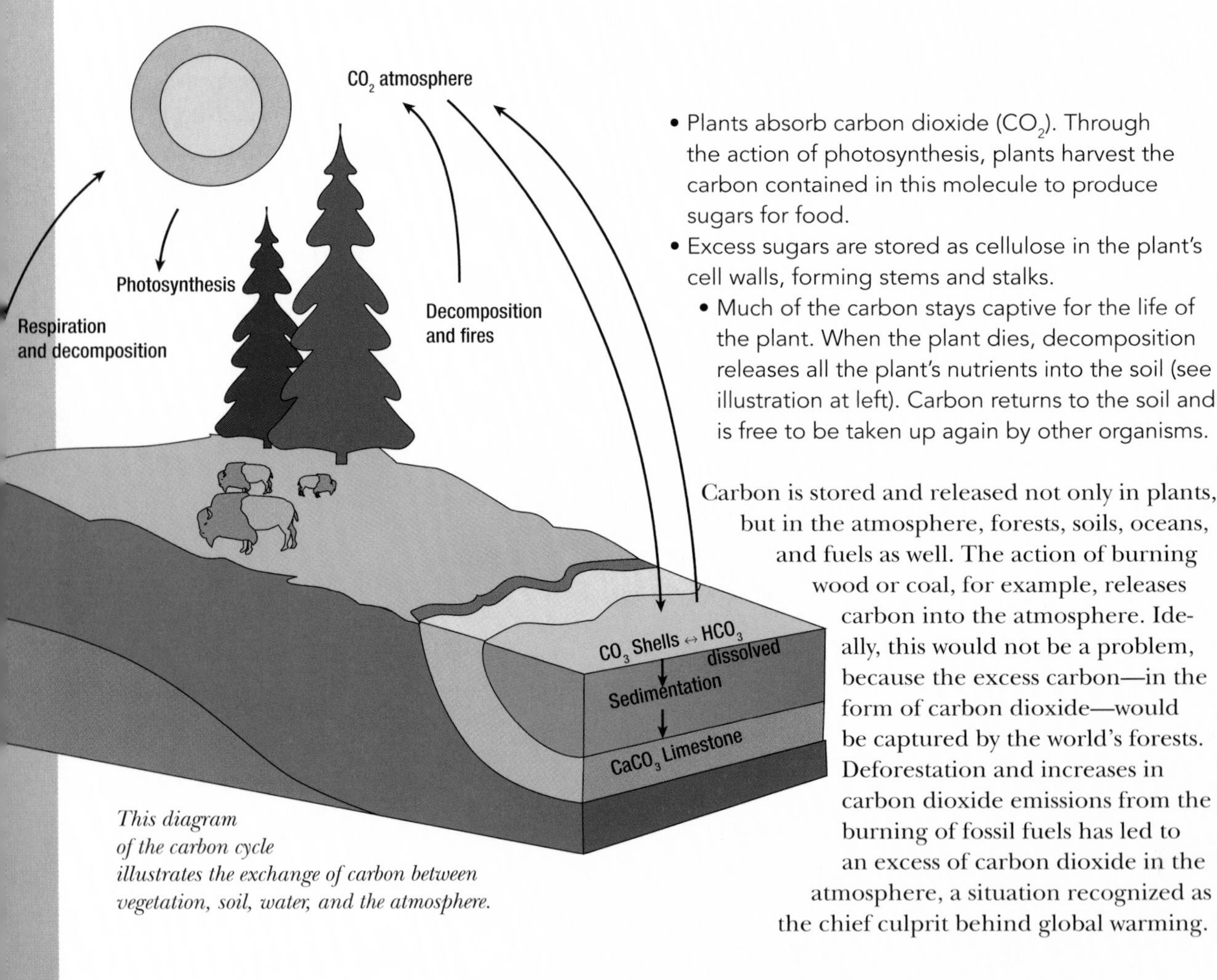

This diagram of the carbon cycle illustrates the exchange of carbon between vegetation, soil, water, and the atmosphere.

- Plants absorb carbon dioxide (CO_2). Through the action of photosynthesis, plants harvest the carbon contained in this molecule to produce sugars for food.
- Excess sugars are stored as cellulose in the plant's cell walls, forming stems and stalks.
  - Much of the carbon stays captive for the life of the plant. When the plant dies, decomposition releases all the plant's nutrients into the soil (see illustration at left). Carbon returns to the soil and is free to be taken up again by other organisms.

Carbon is stored and released not only in plants, but in the atmosphere, forests, soils, oceans, and fuels as well. The action of burning wood or coal, for example, releases carbon into the atmosphere. Ideally, this would not be a problem, because the excess carbon—in the form of carbon dioxide—would be captured by the world's forests. Deforestation and increases in carbon dioxide emissions from the burning of fossil fuels has led to an excess of carbon dioxide in the atmosphere, a situation recognized as the chief culprit behind global warming.

The Nitrogen Cycle

Earth's atmosphere is made up of nearly 80 percent nitrogen gas. Humans and other animals need nitrogen to produce healthy cells. Plants need the element as well. But living organisms cannot always use nitrogen in its natural form. So nature has developed ways to alter nitrogen so it can be more readily absorbed by living things.

- Soil bacteria harvest nitrogen out of the air and convert it to nitrates, a chemical form that is more readily absorbed by plants.
- During storms, lightning causes nitrogen to bond with rainwater, and it falls to the Earth in nitrate form.
- Plants absorb nitrates and thrive.
- They, in turn, are eaten by herbivores, who absorb the nitrates into their bodies.
- Plants and herbivores are consumed by humans, who absorb the nitrogen.
- All living things return nitrogen to the Earth upon excretion and death. A second type of soil bacteria—denitrifying bacteria—breaks down wastes and decaying matter and releases nitrogen back into the atmosphere as gas. The cycle begins anew.

Excess nitrogen in the soil and water—from human-made fertilizers and other pollutants—can be a problem. Excess nitrogen accelerates plant growth, causing rapid consumption and eventual destruction of habitat.

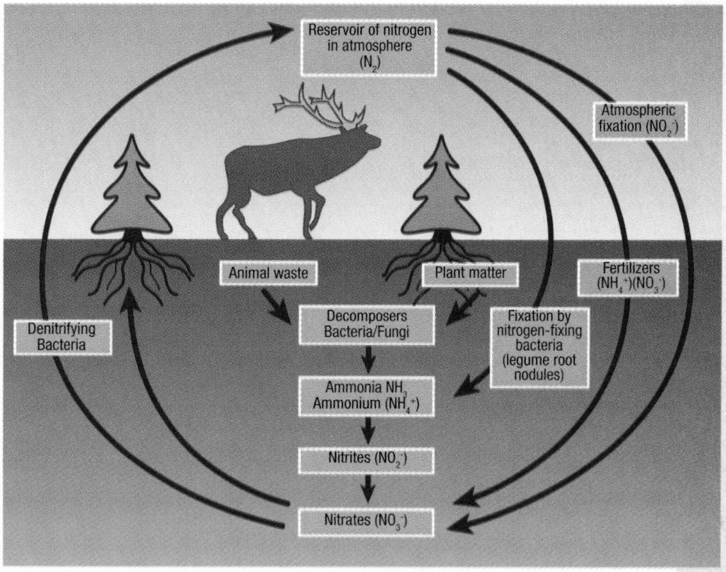

The nitrogen cycle accounts for the transformations that all nitrogen-containing compounds in nature endure.

The Hydrological Cycle

The water cycle is probably the best-known cycle. We humans use water every day, and understand how important it is to our lives. But we probably do not realize that we interact with water on a small part of its journey around Earth's atmosphere.

- The water we drink begins as gas vapor in the atmosphere. The gas condenses in clouds, forming liquid, which falls to Earth as rain or snow.
- Wherever the precipitation falls, it is channeled by rivers and streams to a larger collection area, such as a lake or reservoir. A lot of precipitation ends up in the world's oceans. Some of it seeps underground, where it collects in larger underground aquifers, layers of spongy rock.
- Water that remains in reservoirs or in underground layers remains fresh and can be harvested by humans and used. Water that is polluted by human and industrial use is treated as sewage and returned to the environment.
- Water eventually evaporates back into the atmosphere. Plants that have taken up water through their roots return water to the atmosphere by transpiration through their leaves. Humans and animals also contribute water to the atmosphere as they breathe and sweat.

The water cycle, like the other two, has been severely impacted by human intervention. Rising world temperatures are causing more of Earth's frozen water to be released into the atmosphere. And those same high temperatures are accelerating evaporation. This all adds up to more water being pumped into the cycle, possibly resulting in more extreme weather events. Though this is not accepted by all scientists, the extreme weather trend has been observed by insurance companies, who are forced to pay for the damage associated with extreme weather events.

CHAPTER 7

MATERIALS

Above: Calcium carbonate, an ionic compound abundant in nature, is found in eggshells, seashells, and in dramatic stalagmite formations in limestone caves. Used in many consumer products, such as antacids and chalk, it is also the main ingredient in building materials including cement and marble. Top: Ducks swim in a river polluted by plastic bottles and other virtually indestructible manmade materials. Left: Bubble wrap, made of low-density polyethylene (LDPE), is a versatile, elastic material used to cushion and protect objects during transport.

All matter begins with the atom and its subatomic particles. A single atom, representing one unique element, need just combine with another atom and the chemical personality of the two will change. Throughout the realm of chemical interaction, the bonds that are formed, the number of atoms that are involved, and the conditions under which it all takes place will determine if a material is solid or gas, rough or smooth, malleable or rigid, stable or reactive. The design of many of these different materials has taken place without any help from us for as long as hydrogen has floated about in the cosmos. However, amazing strides in materials science are the result of human interaction with these chemical elements—shaping them, combining them, altering them and, in the process, creating problem-solving materials that touch every aspect of our lives, from the medicines we take to the plastics we cannot break. However, we have also created virtually indestructible materials that threaten to stay with us, in our water and in our landfills, forever. The solution to this quandary will come, again, from chemistry, as we have already seen in the design of plastics that degrade in the presence of sunlight and water.

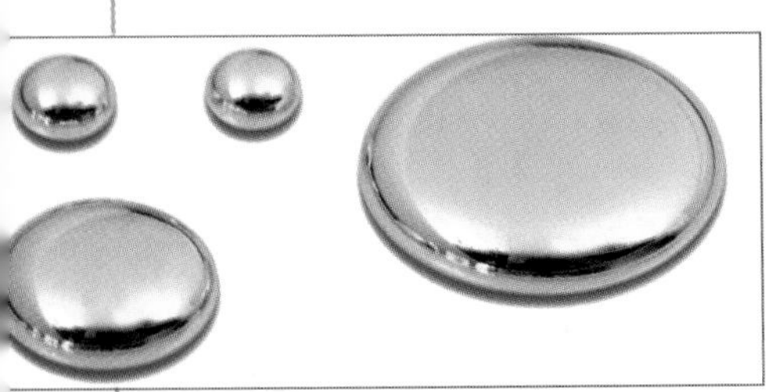

Metals

Metals are the elements we think we know. No one has ever seen a molecule of helium or hydrogen, but we've all admired copper cookware, sported gold jewelry, or eaten with silverware. Based on our familiarity with these shiny, durable objects, many of us are likely to say we definitely know what metals are. But we would be wrong.

Mercury is a metal, and it is far from hard and durable. In fact, it's a liquid at room temperature. Sodium is a metal, and we may gobble thousands of milligrams of this metal every day. Praseodymium is a metal, but how many of us have ever heard of this grayish white material used in airplane manufacturing and specialty eyewear?

Copper, like many other metals, is malleable and ductile. It can be hammered, bent, and stretched into a range of functional formations, including copper pipes (top) and conductive electrical wire (above). Top left: Mercury is a metal that naturally maintains a liquid state at room temperature.

PROPERTIES OF METALS

Scientists are careful to list a number of attributes when defining metals, because these substances may have such enormous superficial differences. There are a few general guidelines that can be used to identify metals, although many of the attributes vary widely in scale.

The single most important attribute metal elements share is that they easily form positive ions and join by metallic bonds. Metals have electrons that are not tied to a single atom but are free to move around, zipping around from one nucleus to another. But when an electron leaves a metal atom, the atom's center becomes positively charged. We call this is a positive ion. If we could visually inspect a group of metal molecules, we would see a dense network of positive ions at the center of each and every atom, surrounded by a cloud or gas of flitting electrons.

Metals also tend to conduct electricity. These free electrons allow metals to transport an electric charge, but that does not make all metals great conductors. That is why we use copper in wires, not tin. One does a better job than the other, but both are conductors. Conductivity decreases when metals are heated, as the increasing temperatures cause more dynamic vibrations in the atoms that disturb the flow of electrons.

Often malleable and ductile, metals can be hammered into new shapes, or stretched finely into wire. This is usually true, but remember our friend mercury? It *flows* like water and resists attempts to be fashioned into wire. If we could visually inspect metallic atoms as we are banging, bending, or hammering them into new shapes, they would look like dense mats of ball bearings rolling against each other but still sticking together.

Many metals are opaque and lustrous, although some hold a shine better than others. (That's probably why we are fonder of gold than lead.) But gold can be stretched out so thin that it can be used for gilding large surfaces. But as thin as a sheet of gold leaf is—only a few micrometers—we still cannot see through it. It is opaque.

About 75 percent of the elements on the periodic table are metals, but this percentage is somewhat misleading. Metals are not nearly as abundant as nonmetals such as hydrogen, helium, and oxygen. In fact, the 86 known metals make up only about 25 percent of the Earth's mass. Yet they are the elements that have most impacted human civilization. Indeed, we characterize the development of human culture according to our ancestors' ability to mine and use metals: The Bronze Age and the Iron Age.

Humans have always displayed a remarkable ability for identifying the precise attribute or attributes of a specific metal and honing in on its usefulness. Centuries ago, we admired gold's malleability, ductility, and resistance to tarnish so much that we prized it for jewelry and stamped it into coins. Slippery mercury found its way into our scientific instruments because it stayed liquid at extreme temperatures. Copper was an excellent conductor of heat, so we used it for our cookware. When we discovered, through painful trial and error, that copper taints food, we lined our copper pots with safe, food-grade tin. To this day, we continue to match a metal's unique properties to a specific type of work we wish to accomplish.

The Mayan pyramid of Kukulkan in Chichen Itza, Mexico.

MATERIALS THROUGH THE AGES

The early history of humankind is often divided into a three-part time scale—the Stone Age, the Bronze Age, and the Iron Age—based on the kinds of tools early peoples used. Scientists reason that tools are a form of technology, and one can judge how sophisticated a culture is by determining when it began using certain tools. Not all cultures adopted these tools at the same time, so it is possible that while one culture was still using stone tools, a nearby or distant culture was already melting copper and tin to make the alloy bronze. The scale is handy, but problematic. Some societies skipped over some of the ages. Others, like the ancient Mayans, became very advanced without ever leaving the Stone Age. And still others kept using stone tools even after they had achieved bronze or iron tool-making. Incidentally, the Iron Age comes last because iron requires more heat to melt than bronze, and was thus more complicated to work with. It is believed that early American cultures never reached the Iron Age because ready-made iron tools were introduced to the continent by European explorers.

Alloys

In its pure state a metallic element is rarely ideal for human commercial or industrial use. To make a raw metallic element more durable or ductile, we mix it with other elements. The result is called an alloy. Just as parents pass their traits along to their child, two or more metals convey a little of themselves to a new alloy. Alloys are almost always stronger than the original elements because when different-size atoms bond together, it becomes more difficult for them to move in concert. The result is stiffer, stronger metal.

Some metals are so unusable in their raw state that they are used only in alloy form, though we still refer to them by their "pure" name.

Gold is one of these. The metal we call gold is often an alloy of pure gold mixed with the elements copper, nickel, silver, palladium, and zinc. Just how much pure gold is present in the metal is expressed by its *carat*. Eighteen-carat gold is 18 parts gold out of 24 total parts metal. Since 18/24 is 3/4, jewelry made from this metal is 75 percent gold. In the same way, 12-carat gold is 12/24, or 50 percent; 14-carat gold is 14/24, or 58 percent; and 24-carat gold is 24/24, or 100 percent. We rarely see 24-carat gold because it's too soft to work with—the very reason for gold alloys in the first place.

Above: In its pure form, gold is so soft and malleable it is virtually unusable. When combined with copper, nickel, silver, palladium, and zinc, a gold alloy becomes stiffer, stronger and can be put to practical use. Top left: Aluminum foil, made from an alloy that contains between 92 and 99 percent aluminum, is prepared in thin sheets and is extremely pliable.

Silver is another alloy that takes its elemental name. When we buy sterling silver, we are really buying an alloy that is 92.5 percent silver and 7.5 percent copper. The copper strengthens the silver just enough to stand up to the rigors of frequent handling. Until the mid-twentieth century, the British monetary unit, the pound, was equal to one troy pound of sterling silver.

Aluminum is yet another popular element that is almost always used in an alloyed state. Why? Because it becomes almost 100 times stronger when mixed with metals such as copper, magnesium, and zinc. Other alloys take new names not found on the periodic table.

Bronze was the first alloy ever made by humans. Originally, bronze was made by mixing tin and copper. Early humans found that the two metals, usually found together in rock, melted easily over flames. Separately, copper and tin are fairly soft. Think how easy it is to bend a child's tin toy or a strand of copper wire. Tin atoms are larger than copper atoms, so when the two are combined, the copper atoms cannot move as easily as they do on their own. The new alloy is harder but also adopts copper's corrosion-resistance.

Solder can have a wide range of compositions, the simplest of which are 60 percent tin and 40 percent lead. It is prized for its low melting point, which allows it to be melted with soldering tools, and to be used like a glue to bond two or more separate sheets of metal.

Steel alloys contain varying percentages of carbon, nickel, manganese, and other metals, depending on the steel's intended use. At this urban construction site, steel will be used throughout the structure being erected—in the connectors, plates, nails, bolts, and screws.

Pewter used to be made of tin, copper, and lead. It looked remarkably like silver but was much cheaper, and so was used to make a wide variety of dinnerware and home decorative items such as candelabras. Pewter's lead content made it easy to craft using hand tools or powered lathes. But lead is poisonous, and its presence no doubt harmed people who ate off pewter plates or drank from pewter tankards. Modern pewter blends tin with copper and safe lead-substitutes such as antimony or bismuth, or both.

In 2002, Jonathan Lee, a NASA structural materials engineer, invented a new high-strength aluminum-silicon alloy that may ultimately lower engine emissions and improve gas mileage in cars, boats, and recreational vehicles.

Steel is the workhorse alloy of many industries. It is mostly iron, with varying percentages of carbon, nickel, manganese, and other metals, all depending on the steel's intended use. Adding the latter elements ensure that the new alloy will be stronger than iron would be on its own. Cast iron is heavy and durable, but brittle. If a cast-iron pot is dropped, for instance, it may crack. If we add chromium to iron and nickel, the chrome forms a surface coating that resists rust and stains. This alloy is stainless steel. Some chefs prefer high-carbon steel knives to stainless steel because the former tend to stay sharp longer.

Ceramics and Glass

Ceramics and glass are as at home in the world of art and decor as they are in a chemistry lab. The touch, feel, and shimmer of an exquisite crystal goblet is often reserved for the most special of occasions. The colors and forms of ceramic art sculpture have captivated the human eye since before the reddish clay of ancient Egypt was molded and baked, creating ceramics that were used for jewelry and urns. The history of uses of ceramics and glass is a long one, ranging from the plumbing of ancient Rome to the latest in fashionable eyewear. The numerous future applications in the chemical industry for both of these versatile materials hold promise in fields ranging from medicine to telecommunications.

Above: Heating is an important part of the glass-making process, but it is during the cooling phase of fabrication that glass takes on its unique qualities—strength, resistance to chemical corrosion, and extreme durability. Top left: Stained glass began to flourish as an art during the tenth century, when metal oxides were used to add color to molten glass heated in clay pots over a furnace. Copper oxides were added to produce green, cobalt for blue, and gold to produce red.

STATE OF MATTER

Glass in its natural form has always existed, and can occur when lightning strikes or a volcano erupts, heating minerals to the melting point and then forming when they cool down. The molten action of volcanoes is a major factory for obsidian, a kind of natural glass that is believed to have been used since man decided to start making his own tools.

While heating is an important part of the glass-making process, it is during cooling that glass takes on its unique state. The most common ingredient in glass—though glass can be manufactured from a wide variety of materials—is silica, which is a major component of sand. When silica or another crystalline material is heated to very high temperatures and then cooled rapidly, the atoms do not have time to reform into the orderly rank-and-file crystals that were previously present. Rather, when glass cools the atoms are captured in a fairly random, disorganized, more free-flowing arrangement, more similar to a liquid, though the result is solid. So while glass shares properties with both solids and liquids, it is different from both and better described as a material state rather than a specific class of

In ancient times, obsidian—a type of naturally occurring glass produced when lava cools quickly and freezes—was used to make arrowheads. Today, some modern surgeons opt to use obsidian surgical scalpels because their blades can be sharper and thinner than those of traditional steel instruments.

material. It is firm like a solid with the lack of a locked atomic order reminiscent of a liquid. Glass, once formed, is very strong, resistant to chemical corrosion, and can last more than an estimated one million years in a landfill before it begins to break down.

Ceramics is a broad term that encompasses a wide variety of inorganic and nonmetallic materials that usually require the application of heat to solidify. Pottery, often used interchangeably with ceramics, is made primarily from clay and in the heating process does not entirely melt to the liquid state. One of the main differences between ceramics and glass lie in their bond structure. The covalent bonds prevalent in glass are not as prevalent in ceramics, which more often contain ionic bonds that can form more distinct and structured crystals. Glass-ceramics, on the other hand, are made using a process similar to what is used to create glass. However, the material is heated to even higher temperature. Part of the material takes on a more uniform, crystalline pattern. If this crystalline pattern is present in more that half of the material, it is considered to be glass-ceramic. Glass-ceramics are both harder and stronger than glass.

The importance of glass and ceramics to chemistry, chemical research, and consumer products could fill several books on its own. Ceramic tiles are used to protect the space shuttle from heat damage. Fine optical glass fibers as strong as glass can send data for miles and are changing the speed at which the world can communicate. Fiber optics make it possible to transmit data via laser light. Glass-ceramic cooktops can be found in the most high-tech of kitchens. And minuscule, radioactive glass beads are being sent into the human body as part of groundbreaking new treatments for liver cancer.

Induction cooktops conserve energy by transferring heat directly into cooking vessels.

GLASS CERAMICS GET COOKING

Cooking can be fun, but also dangerous and time-consuming. People can burn themselves and their food, and have a hard time cleaning their blackened ranges. Induction cooktops can help. They are about 50 percent more efficient than other cooking surfaces. To work, they use the principle of induction: When iron molecules are placed within strong and rapidly changing electromagnetic fields, their electrons vibrate and heat up. The surface of an induction cooktop is a glass ceramic sheet. Under the ceramic are flat copper coils. When the range is on, the coils emanate electromagnetic waves. The waves flow through the ceramic, and only the pot and the food in it get hot. Because the ceramic stays cool, cooks cannot burn their hands if they touch it. Cleanup is easier because food cannot burn on the glass ceramic. After dinner, the cook wipes the ceramic down with a wet cloth. The downside: Only ferrous—iron-bearing—alloys work on induction ranges. Stainless steel and cast iron pots are fine, but copper, aluminum, and ceramic are not.

The Chemistry of Carbon

Carbon brings to mind a variety of images: Carbon copies come to mind, as do chunks of charcoal in an outdoor grill, the dangers of carbon monoxide that we so often hear about, and of course the carbon dioxide we exhale all day, every day. And while carbon is a part of all of these, it is also the second most abundant element in the human body, coming in behind oxygen, and is the sixth most abundant element in the universe. From DNA to diamonds, carbon's uses in its natural state, as well as its many industrial applications, make the study of this element and its many compounds an endless pursuit.

THE BUILDING BLOCK OF LIFE

Carbon occurs naturally in various forms or allotropes. The most familiar are diamond, graphite, and amorphous carbon. Part of the reason for carbon's incredible versatility lies in the structure of the atom itself. Carbon has an atomic number of six, meaning it has six protons and six electrons. Those six electrons orbit the nucleus in specific areas referred to as shells or clouds. Carbon's six electrons occupy two shells: Two electrons are in the shell closest to the nucleus, and the other four electrons are in the outer shell, known as the valence shell. Because the valence shell has the capacity to hold eight electrons, carbon has four empty "parking spaces," so to speak, which means that four other atoms can pull into those spaces. The ability of carbon to readily bond with four other atoms is an important part of the chemistry of the element. But it is also carbon's ability to bond with itself that gives it the capability to form molecules consisting of long chains of carbon. This makes carbon a key constructive element, allowing it to form the backbone of many complex compounds in the human body. Photosynthesis, the reaction that changes sunlight, water, and carbon dioxide into sugar and oxygen, has carbon

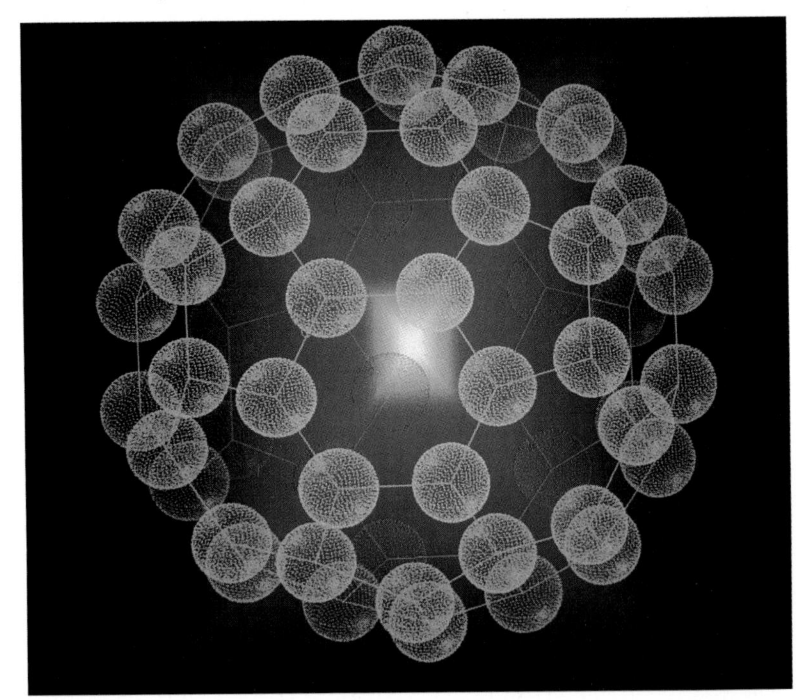

Above: In this computer-generated image of a buckyball (buckminsterfullerene, C60) molecule, the green spheres represent carbon atoms and the lines signify the bonds between them. Discovered in 1985, buckyball is a fullerene—an allotrope of carbon—with physical and chemical properties that can be exploited to produce new catalysts, lubricants, and superconductors. Top left: Diamonds, a naturally occurring form of carbon, are the hardest natural substances known.

Charcoal is dark or black porous carbon, often produced by heating wood in the absence of oxygen.

compounds as both a reactant (carbon dioxide) and a product (glucose). The carbohydrates that give energy to the human body consist of carbon chains. In short, all plants and animals need carbon to survive, earning it the nickname "element of life."

BEYOND THE BODY

Carbon's uses in industry run the gamut from transportation and life-saving to fashion and face-lifts. It is in medicines, food, clothes, and fuel. Perhaps no other naturally occurring form of carbon communicates the vast range of uses for this element than diamonds. Carbon in diamond form is not just a girl's best friend, but exhibits uses that far exceed mere personal adornment. The hardest natural substance known, diamonds form the stylus on record players and coat knives and drill bits, giving them an incomparable hardness and durability. Diamonds are in tools that help cut asphalt during highway construction, and are ground up and put in face creams to serve as exfoliants. Forms of carbon are used in scalpels, computer chips, and semiconductors—new uses are being discovered and developed constantly. Fullerene, an allotrope of carbon that is composed of 60 carbon atoms arranged in a sphere, was originally discovered in 1985 has come into its own only in the last decade or so. Known by the full name buckminsterfullerene—after Buckminster Fuller, inventor of the geodesic dome—these "buckyballs" exist in the world of optics and semiconductors and present remarkable potential in the area of nano-medical applications as a possible drug delivery system in the treatment of disease. In one particular exciting development, these carbon structures, paired with antibodies, may someday be used to target cancer cells.

Large Hadron Collider located in the facility of the European Council for Nuclear Research (CERN). This portion of the semiconductor tracker for ATLAS, a detector in a particle accelerator, consists of nine carbon-fiber disks, each with 1.5 million thin silicon strip detectors.

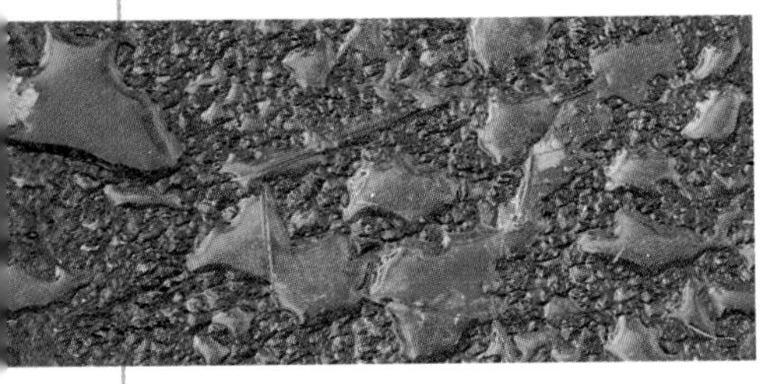

Organic Chemicals

Carbon and the many compounds that it is capable of forming are so important that an entire branch of chemistry is based on it. Organic chemistry examines the structures, and reactions, of organic compounds. The word "organic" is tossed around a lot in modern times with the increased presence of organic foods and a heightened awareness of the origins of the fuels that go into our bodies as well as our cars. Carbon is, of course, a part of both of these fuels—as prevalent in the gasoline put in a car as it is in the sugar put in a cup of coffee. But organic compounds are present in all kinds of foods—including food additives and a wide variety of food sources that are not considered in any way natural, wholesome, or healthy. The first use of the word "organic" was in the early nineteenth century, and it was meant to differentiate between compounds made from living things and those that were not. Of course today organic compounds are found and manufactured without live plants and animals. Still, "organic" is associated with those things that are living, and carbon is an essential element of all living things.

Above: In this computer model of buckyballs (buckminsterfullerene, C60) molecules in a carbon nanotube, the colored peaks show electron waves. The nanotube, a cylinder of carbon atoms less than a billionth of the width of a human hair, containing buckyballs is a good conductor of heat and electricity. Top left: Gasoline, or petroleum, is an organic compound because it contains hydrocarbon chains.

CHAIN GANG

An organic compound is one that contains a carbon-hydrogen bond. Carbon is present in so many compounds that the number of organic compounds is virtually incomprehensible. Carbon can bond to four other atoms at the same time and can create double and triple bonds as well. Whether carbon bonds to itself or is bonded to an atom of a different element, it has the ability to do so covalently. Generally speaking, a covalent bond is a bond in which both atoms share electrons nearly equally. Overall, the bonds that carbon forms are very stable. A major part of the organic chemistry industry—as well as a key function in living beings—is being able to synthesize large organic molecules. In this respect carbon is, in many ways, in a class of its own. There appears to be no size limit whatsoever to the molecules that carbon can make, especially since, in addition to making chains that are literally thousands of atoms long, carbon can also form rings, branching chains, and more. The length and structure of the chains have a great impact on the properties of the organic compound. In our bodies, carbon chains build fats, sugars, DNA, proteins, and much more.

MILLIONS OF YEARS IN THE MAKING

The majority of organic chemicals that are encountered on a daily basis by the average consumer are derived from petroleum. These organic chemicals are aptly referred to as petrochemicals, and they crop up in clothing as often as they do in car engines. Plastics, hand lotions, contact lenses, and antihistamines are just a few of the myriad products made from petrochemicals. Petroleum itself comes from deep inside the Earth where microscopic marine animals have been pressed into the sediment. After being subjected to high-pressure heat for millions of years, they are transformed into oil and gas. There are hundreds of different organic chemicals in crude oil. The refining process, known as fractional distillation, separates these different chemical components by heating them. The difference in boiling points among the organic chemicals separates out the lighter compounds, such as propane, from heavier compounds with longer chains, such as kerosene.

IN THE FIELD

Organic chemists work in a highly creative capacity, not unlike architects or designers. Though organic chemists may spend time learning about existing organic compounds, they also design new compounds—based on the chemical characteristics and properties they seek—and then work diligently to synthesize these creations. Some of the molecules and macromolecules they design and ultimately synthesize are so complex that they rely on computers capable of performing multilayered 3-D modeling.

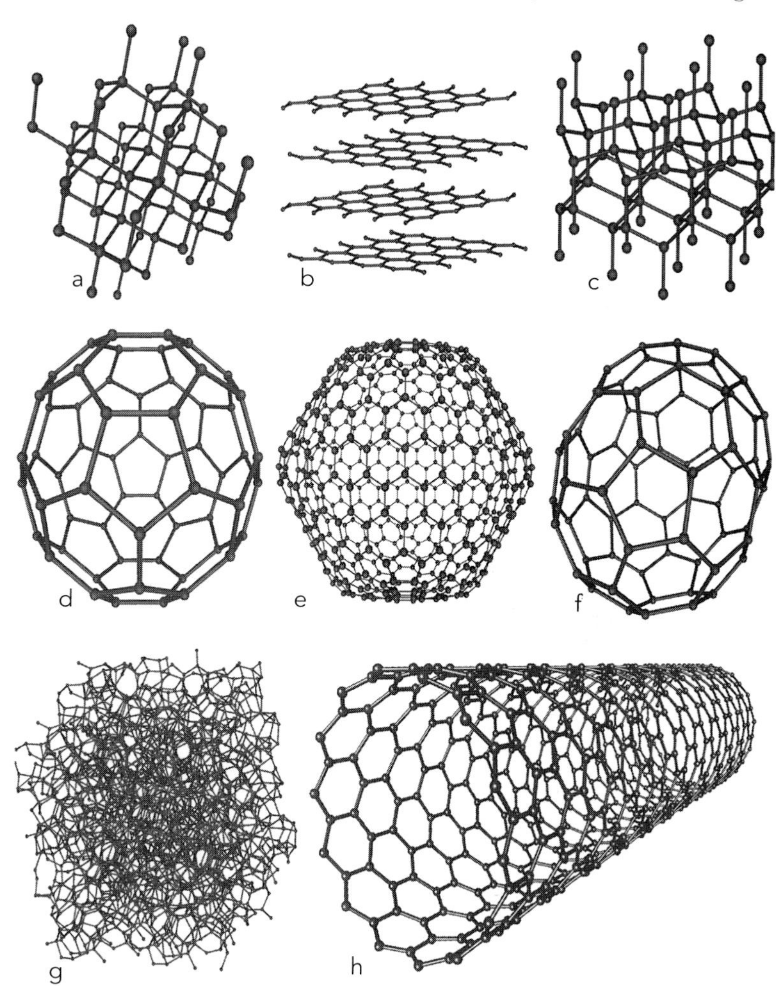

The illustration shows eight allotropes of carbon with different molecular configurations. Diamonds (a) are the hardest known natural mineral; graphite (b) is one of the softest known substances. Lonsdaleite (c) is transparent brownish-yellow in color and referred to as a "corrupted" diamond or "hexagonal diamond." Buckminsterfullerene (C60) or buckyballs (d) are used in optics and semiconductors; C540 (e) and C70 (f) are fullerenes (part of the same family as the buckyball molecule). Amorphous carbon (g) is a glassy substance. Single-walled carbon nanotubes (h) are tiny tubular structures.

Polymers

A trip to the local recycling center is all one needs to learn something about polymers. Dumped in the plastics bin are hundreds, possibly thousands, of bottles, jugs, trays, and other cast-off containers made from these very useful molecules. Our homes, too, are filled with polymer-based products. Soda bottles and sandwich and freezer bags are made from polyethylene. CD jewel cases are made from polystyrene. Our plumbing pipes may be made of polyvinyl chloride. The bullet-resistant glass at our local bank is made of polycarbonate.

The word "polymer" comes from the Greek: *poly meros*, meaning "many parts." Polymers are large molecules that consist of long chains of repeating structures. The repeating structures are called monomers, and they are held together by covalent bonds, which means each monomer shares electrons with the next monomer in the chain. The monomers bond during a chemical reaction called polymerization. Carbon is commonly found at the heart of polymer chains, because it is one of the few elements that can bond four different directions with four other atoms. If we can imagine a polymer as a long charm bracelet, carbon makes up each of the links in the bracelet; the charms dangling from those links are other elements.

By this definition, plastics are polymers, but they're not the only ones. Humans and animals depend on polymers for their very existence. The DNA molecule, the central building block of life, is a polymer, as is a long strand of protein made of smaller amino acids. The sap that flows in trees and the food we eat are replete with polymers.

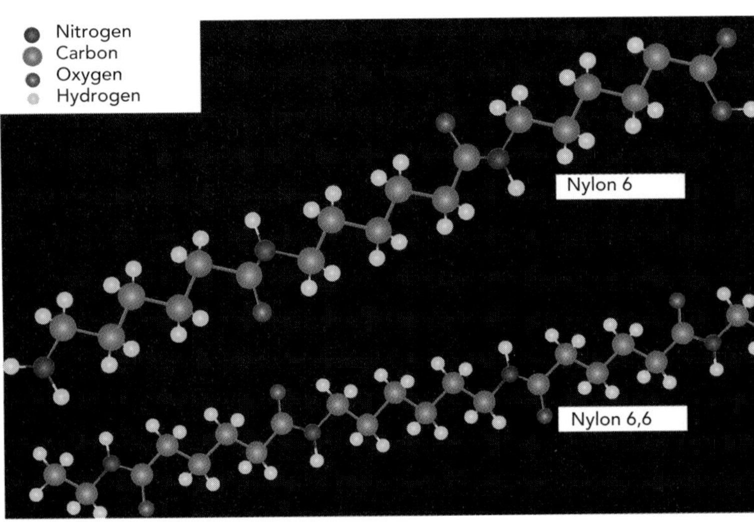

Above: Computer-generated diagram of nylon 6 and nylon 6,6. Nylon-bristled toothbrushes were the first products manufactured using this pioneer polymer. Top left: Plastic polymers vary widely in their composition, durability, and flexibility. These versatile molecules can be made into everything from bowling balls and football helmets to plastic containers and foam mattresses.

THE FIRST MANMADE POLYMER

The first synthetic polymer was nylon, produced in 1935 by the DuPont chemical company as a substitute for silk. It was mass-produced and marketed as a women's stocking material, though the same plastic molecules can be molded into materials such as rope and textiles. Other synthetic polymers can be fashioned into rubberlike substances, liquids or semiliquids that dry into stronger resins or glues, rigid meat trays found in supermarkets, and thin, throw-away plastic films that covers the food in those trays.

Despite their usefulness, synthetic polymers are plagued

Above: A man operates an extrusion machine producing recycled polyethylene sheet tubing. The plastic tubing is blown by hot air that shapes and dries the tube before folding and rolling it. Top: Originally carved from wood, surfboards are now made from glass fiber and polystyrene foams.

by environmental issues. Since they are designed to be durable and degrade only in the presence of other chemicals, high heat, or prolonged exposure to light, these products do not disintegrate readily. A plastic soda bottle may last forever.

As a result, citizens all over the world are pursuing various ways of managing our use of household containers. The choices are clear: We can reduce the amount of plastic containers we use. We can reuse containers for purposes other than their original use. We can recycle the containers we use and fabricate them into useful new products.

PLASTIC FROM CORN

Another approach is to design new polymers that break down more readily, as in the case of a new corn-based product, polylactide, or PLA. To make it, common corn syrup is fermented to make lactic acid, the sour substance in spoiled milk. The acid is heated to boil off water, forcing the acid molecules to bond as doughnut-shaped structures called lactide monomers. Next, a catalyst breaks these bonds and sets off a chain reaction. As each ring pops open, it seeks stability by glomming onto another monomer, forcing *it* to pop open. The process continues, until thousands of monomers make one polymer strand. The strands entwine with adjacent strands, forming tough stringy fibers. The result is a translucent plastic pellet that can be melted or formed into any shape.

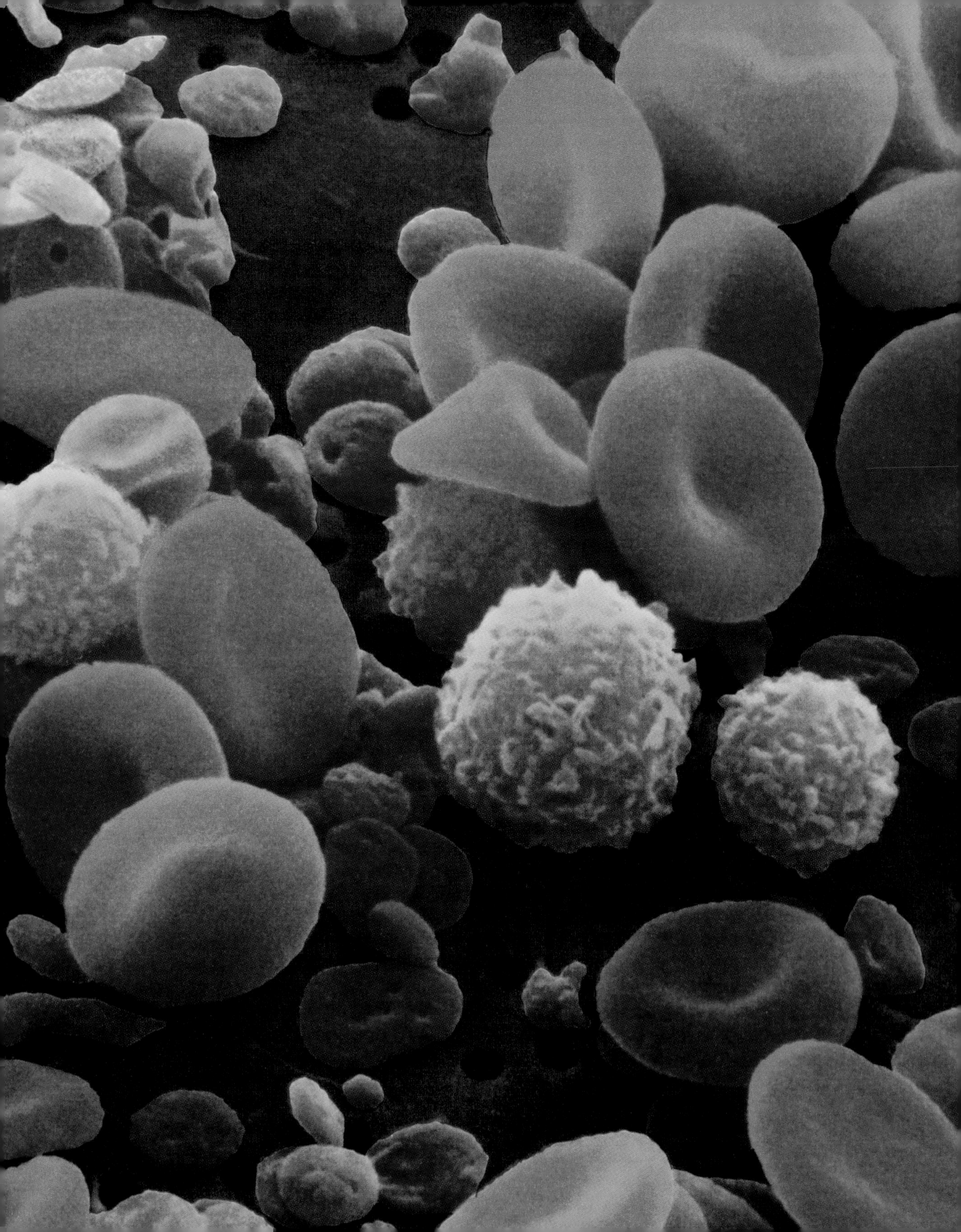

CHAPTER 8

CHEMISTRY OF LIFE

Left: Human blood cells and platelets seen in a colored scanning electron micrograph (SEM). Every living organism, including our own blood and organs, is full of chemicals; some of these chemicals are naturally occurring, others are introduced from outside our bodies. Top: A close-up of the flower Hypericum perforatum, *or St. John's wort, a plant that contains chemical compounds believed to be beneficial in the treatment of depression. Bottom: A worker in a pharmaceutical laboratory.*

The perception that the natural world and the chemical world are separate and distinct is as easy to subscribe to as it is to dispel. Something that is "natural" is thought to be void of "chemicals." Chemicals are not merely in the world around us, but are pulsing through our veins, carrying vital information to our organs, and maintaining our homeostasis under extreme or even dangerous conditions. Some of the chemicals in our bodies are constantly and successfully fighting germs and diseases we will never even know we had, while others are working overtime to convert the food consumed into a usable form for our bodies. But there is always a tipping point. The body can also be attacked and worn down by chemicals. Some of these synthetically produced chemicals were created in an effort to make our lives easier and more productive. Others are naturally occurring yet can still harm us when they are present in the wrong amounts in an organism seeking balance. Still, chemical research is relying more than ever on natural resources and wisdom for methods that can reduce toxins and their effects on organisms living in a modern world.

Fundamental Chemicals of Living Things

In order to be bowled over by the awesome spectacle of chemistry at work, all humans need do is take a good look within. The chemistry that surrounds us is no more fascinating than the chemical reactions that are constantly at work within our bodies. The chemistry of all living creatures, whether microscopic or gargantuan, is greatly dependent on carbon-containing organic compounds. There are certainly nonorganic compounds—breathable oxygen and water, to name just two—that play a vital role in the day-to-day functioning of living systems, and every molecule makes its own life-sustaining contribution. However, the preponderance of certain key, fundamental chemical compounds in living things merits much closer examination.

Above: Bread is one of many foods that contain carbohydrates, one of the four organic compounds found in most living things. Top left: A three-dimensional model of a sugar molecule.

THE FANTASTIC FOUR

There are four predominant classes of organic compounds that are found in the majority of living things: carbohydrates, lipids, proteins, and nucleic acids. These are also referred to as biomolecules. Each one of these plays a major role in keeping organisms running smoothly, giving them the ability to make, process, utilize, and store energy as well as reproduce and propagate their species.

CARBOHYDRATES

Carbohydrates are the energy stores, the back-up generators for living things. Often referred to in general discussions about diet, carbohydrates are commonly associated with foods such as pasta and bread. While these food items do contain carbohydrates, table sugar is also a carbohydrate. In fact, carbohydrates are just that—chains of sugars. They can be very short or complex and are categorized in a very logical fashion according to the number of individual sugars that they contain: monosaccharides contain one, disaccharides contain two, trisaccharides, three, and when several of these carbohydrates combine together the result is called a polysaccharide. Not only do humans and other animals rely on carbohydrates as a source of energy, key structural components of many marine invertebrates contain another form of carbohydrate, cellulose. Cellulose, classified as a polysaccharide, is also one of the key structural elements in the cell walls of plants.

LIPIDS

Although lipids are essentially thought of as fats, that is only part of the story. The larger class of lipids is primarily composed of fats and oils, and lipids are certainly used as a means of energy storage in living things, including humans. However, there are many different kinds of lipids, and this class of organic compounds also includes steroids, waxes and phospholipids, and glycolipids. The primary trait of all lipids is that they are soluble in organic solvents but insoluble in water. Triglycerides, which are made up of glycerol molecules bonded to fatty acids, are lipids and make up most of the fat in the foods we eat. Whether a fat is saturated, monounsaturated, or polyunsaturated depends on the number of double bonds present—none,

Above: Eggs contain protein, found in the white, and lipids, or fats, found in the yolk. Top: Olive oil is a lipid, an organic compound that is insoluble in water.

one, and more than one, respectively—in the fatty acids. When glucose (stored as glycogen) is running low, lipids step up to the plate. One example of lipids beyond the realm of fats can be seen in the waxy surface on the leaves of some plants.

PROTEINS

The human body is composed of least 50 percent protein, which is the main component in cells. The number and different kinds of proteins present in our bodies cannot be counted, but form everything from the skin on the outside of our bodies to chemicals like insulin that do their work all day long chugging through our veins—proteins are, literally, everywhere. Much in the same way that carbohydrates are made up of chains of simple sugars, proteins are made up of chains of amino acids. In the natural world, there are 20 basic amino acids that are combined to make all the proteins we need. While the human body is capable of making roughly half of these 20 basic amino acids necessary for healthy living, the others have to be obtained through the ingestion of food.

NUCLEIC ACIDS

Nucleic acids in the form of DNA and RNA help create the blueprints, the design for life. The information carried in DNA and RNA makes each and every one of us unique. Again, similar to carbohydrates and proteins, nucleic acids are also made up of chains of smaller units. In this case, the smaller units that combine to make nucleic acids are called nucleotides. The basic nucleotides contain a sugar, a hydrogen phosphate, and one of five nitrogenous bases: guanine, adenine, thymine, cytosine, or uracil.

Athletes sometimes engage in "carbo-loading" in an effort to boost glycogen in the muscles.

CARBO-LOADING

The term "carbo-loading," often heard in athletic circles, is a competitive strategy that involves eating large amounts of carbohydrates or "carbs" before a big race as a means of beefing up the body's energy supply. The aim is to increase the amount of glycogen in the muscles. Glycogen is a polysaccharide (meaning "many sugars"), and when excess glucose is present in the body, it is stored as glycogen that then functions as an energy reserve for the muscles. The original idea behind carbo-loading, which has been around roughly 40 years, was that by first depleting the body's carbohydrate intake, an athlete would be more likely to store excess glycogen when he or she began eating carbs again. This evolved into another technique that simply involves eating more carbohydrates and reducing training three days before an event.

Photosynthesis

Whenever we bite into an apple or another piece of fruit, we are getting a watery mouthful of natural sugar, or fructose. The same is true when we chomp into vegetables. Some, like sugar beets, which contain lots of sucrose, another natural sugar, are more obviously sugary than others, and others are starchier. Starch is a complex carbohydrate that is related to the various types of sugars found in plants. So, based on our experiences with these kinds of foods, we should not find it too surprising that plants manufacture their own sugars. In fact, this is the job of plants. All day long, they make sugar to nurture themselves, to build cell walls, and to repair themselves. A plant makes millions of glucose molecules in a second, more than it can ever use immediately. It stores the extra glucose as starch in its stems.

How do plants make these delicious sugars? They do it through a fascinating process called photosynthesis, which is perhaps the most important chemical reaction on Earth, because plants are not the only ones that benefit from it. Humans and animals also eat the food produced by plants, and we breathe the leftover oxygen exhaled by plants as a

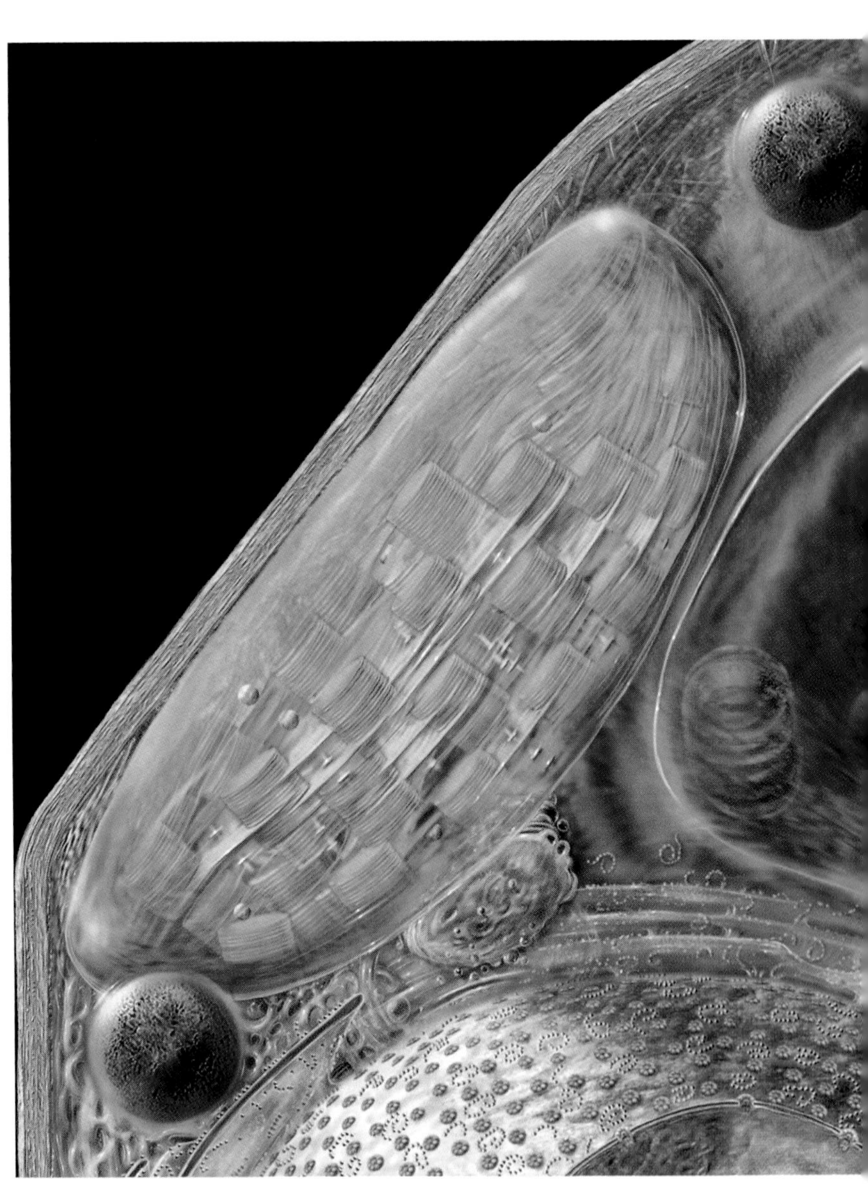

Above: Artwork showing a chloroplast, the part of plant cells that contains the chemical chlorophyll. Chlorophyll allows the plant to convert water, carbon dioxide, and sunlight into carbohydrates, through a process called photosynthesis. Top left: Apples and other fruits contain fructose, a natural plant sugar.

by-product of photosynthesis. All life on Earth sits squarely on the leafy shoulders of plants. If plants ceased to exist tomorrow, all other life would as well. That's why plants, big trees in particular, are often called "the lungs of the earth."

A COMPLEX CHEMICAL REACTION

Photosynthesis is a complex chemical reaction, and some parts of it are still not completely understood. But in general, when a plant absorbs water, carbon dioxide, and sunlight it produces sugar molecules and expels oxygen and some water vapor. The action of photosynthesis happens in the leaves of the plant, involving a chemical called chlorophyll. This is the chemical pigment that gives plants their green color. The job of any pigment is to absorb light; chlorophyll absorbs light from the red and blue bands of the visible light spectrum, but does not absorb green. That green is instead reflected out of the plant, which is what we see.

Scientists consider photosynthesis to be a process composed of two distinct stages: photosynthesis I and photosynthesis II. Photosynthesis I can only take place in the presence of sunlight, and is thus termed the "light reaction." When sunlight hits a plant's leaf, electrons in the chlorophyll molecules become excited and snap off, setting off a chain reaction that produces two high-energy molecules called adenosine triphosphate (ATP) and nicotinamide adenine dinucleotide phosphate hydrogen (NADPH). These two molecules work together during the photosynthesis II stage to break down carbon dioxide molecules and synthesize a precursor to the glucose molecule. This second stage does not need to take place in the presence of light. Historically, it was called the "dark reaction." Today it is more commonly called "light-independent."

What goes on in a leaf is not unlike the constant flow of electricity that runs through a photovoltaic cell when the Sun is shining. In each setting—the solar panel or the leaf—the Sun stimulates the flow of electrons that generate a current, which performs some work. In a solar panel, the current powers homes, shines lights, or runs appliances. In the leaf, the current produces edible foodstuffs, which is either consumed by the plant or stored for future use.

Beautiful fall color is the result of the breakdown of chlorophyll in leaves, as days grow shorter and nights grow colder.

AUTUMN COLOR

The sight of arresting red and gold leaves hanging from the boughs of maples and oaks in the fall is one of the great thrills of the changing seasons. The colors look so vibrant and alive, but actually the deciduous trees are getting ready to slow down for the winter months. The chlorophyll in leaves, used in photosynthesis, gives them their green color. But there are other natural pigments always present as well. When nights get chillier and days get shorter, chlorophyll begins to break down and the green color that once masked the other pigments begins to disappear. This allows the other colors to be fully expressed in a spectacular, if fleeting, display.

DNA and RNA

The master code for every human being resides in a complex molecule in the nuclei of cells called deoxyribonucleic acid (DNA). The structure of the DNA molecule was first explained by James D. Watson (b. 1928) and Francis Crick (1916–2004) in 1953. They showed that the "macromolecule" adopts the shape of a double helix. Imagine a flexible, rubbery ladder that has been twisted into a spiral shape. The sides of the ladder consist of alternating molecules of the sugar deoxyribose and phosphates. Each rung of the ladder is made of a pair of nitrogen bases. There are only four possible bases: adenine (A), thymine (T), guanine (G), and cytosine (C). These bases pair up exactly the same way every time: A always links with T; G always links with C.

RECIPE FOR LIFE

The DNA carries the blueprint for building everything in the body. Certain sections of the molecule are used to make specific proteins that carry out cellular functions. These sections are called genes. We inherit our genes from our parents. DNA comes packaged inside 46 chromosomes. We get 23 chromosomes from our father, and another 23 from our mother. During the course of our lives, the DNA template helps to ensure that we are constantly repairing our bodies and nurturing ourselves in exactly the same way, every time. To do this accurately, the DNA must be divided when new cells are made, and it must impart its recipes for protein making.

THE ART OF REPLICATION

To make a carbon copy of DNA, machinery inside the nucleus unzips DNA molecules from one

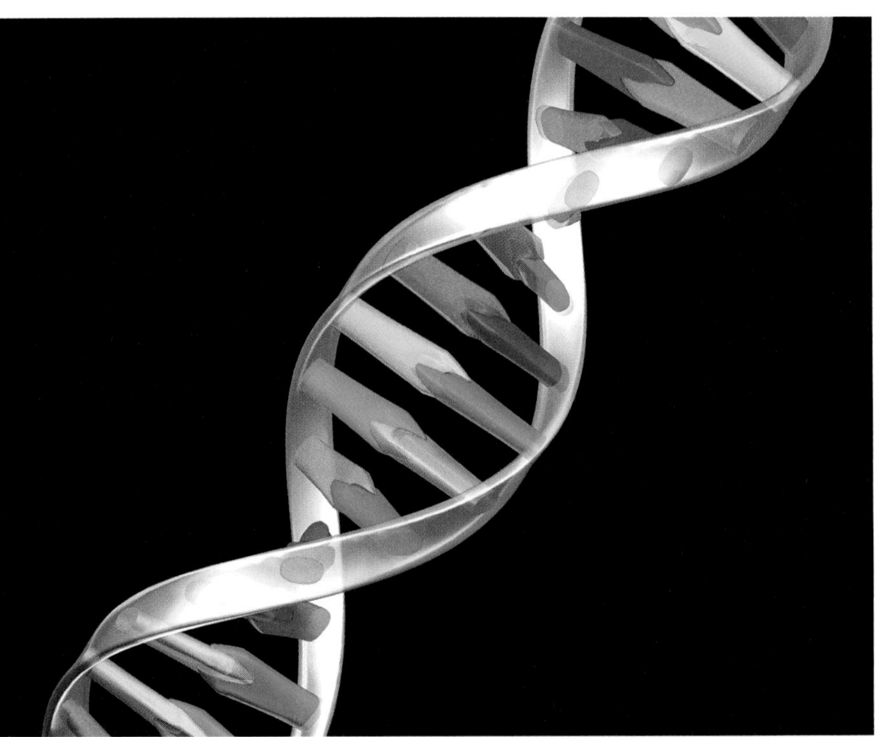

Above: Computer artwork showing DNA. The two white strands are sugar phosphates, which form a double helix, or spiral. The colored rods represent the four nitrogen bases: adenine, thymine, guanine, and cytosine. The bases are always paired with the same match, so adenine (red) always pairs with thymine (blue), while guanine (green) always pairs with cytosine (yellow). Top left: Drawing of a ribosome, a protein-making form of RNA (ribonucleic acid).

end. The rungs of the ladder are now split, and anxious to pair up with their customary nitrogen base. If a broken rung ends with an A, the DNA-making machinery will insert a free T floating around in the nucleus of the cell. The pairs come together, and fresh sugar-phosphate sides are assembled to enclose them. In this way, the DNA molecule is completely replicated as it is unraveled.

MAKING PROTEINS

DNA is like the foreman on a construction job, carrying the blueprints. The proteins are the workers who carry out those orders. In order to make the proteins, the DNA must have a go-between to carry instructions from inside the nucleus to the outside world of the cell.

The DNA's emissary is ribonucleic acid, made of a slightly different sugar called ribose. It is disposable and comes in three types. Messenger RNA (mRNA) is a copy of a gene, or small section of DNA, needed to make a certain protein. The mRNA is like a flash drive that carries information from a main computer to one's laptop or home office. The gene is transcribed exactly, with one exception: For the T base, a related base called uracil (U) is substituted. The newlymade mRNA then migrates out of the nucleus. Once outside, mRNA is enveloped by a ribosome, a kind of protein-making factory that is assembled largely from ribosomal RNA (rRNA), the second type. Transfer RNA (tRNA), the third type, drags free amino acids to the ribosome. During translation, the ribosome builds a protein by putting the amino acids in order according to the instructions on the strand of mRNA. When each amino acid is in place, the finished protein detaches from the ribosome and hurries to do its job. In this way a magnificent hierarchy of tools perform the important work of cellular repair and bodily function.

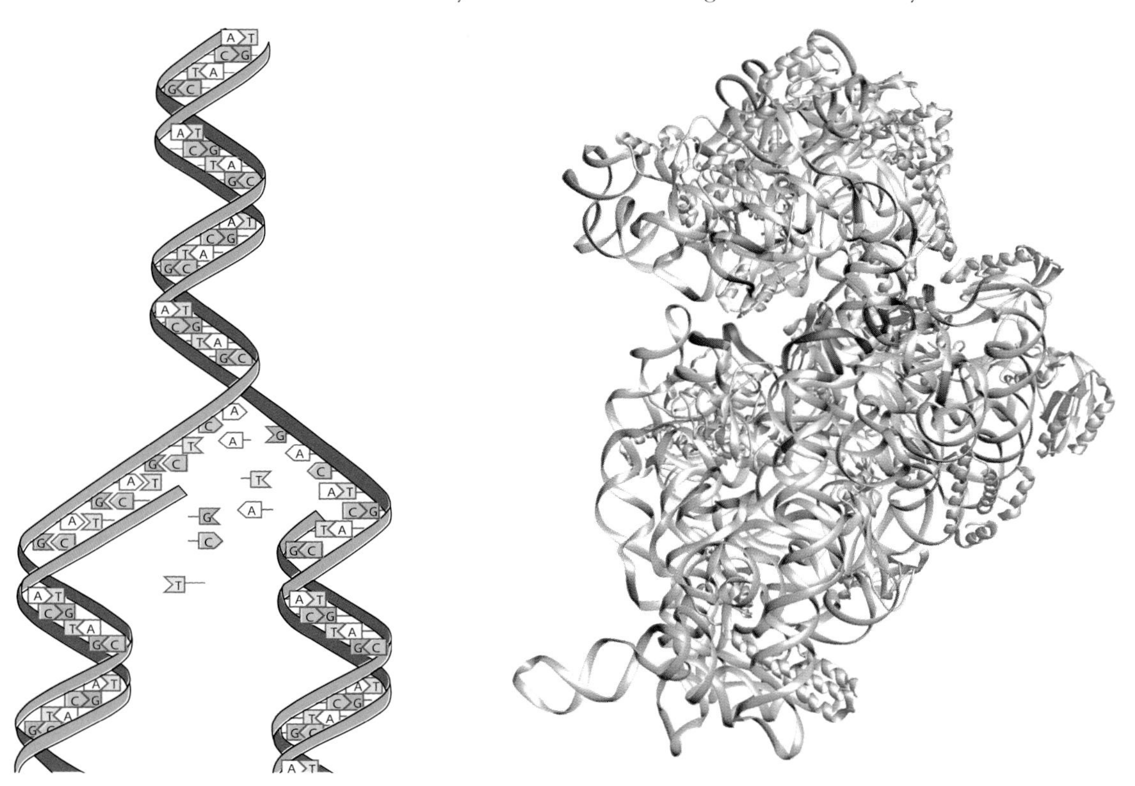

Left: Diagram of a DNA molecule in the process of "unzipping," leaving floating nitrogen bases that will quickly pair up in exactly the same sequence to form a new, perfect copy of the DNA molecule. Right: A ribosome shown in computer artwork.

Chemistry in Cells

Every major activity of a living organism—raising an arm, digesting food, breathing—involves chemistry at the cellular level. The study of these reactions and interactions in living things is called biochemistry. It is, rather simply, the chemistry of living things as seen from the molecular level. The field of biochemistry is vast and comprises everything from how we process the food we eat to how organisms pass genetic material on to future generations.

BREAKING IT DOWN

The molecules that make up the living matter we see around us—our bodies, the plants we grow, animals, and more—are often called biomolecules. It follows that the reactions in which these molecules are involved are referred to as biomolecular reactions. These molecules that form the basis of life are constantly being made and constantly being broken down. The making of these molecules is called anabolism, and their breaking down is called catabolism. The sum total of all these anabolic and catabolic activities is called metabolism.

Metabolism is often referred to with regard to how human bodies break down food, specifically. But it refers to the total of *all* biomolecular reactions, including those involving the building of proteins from amino acids as well as their breaking down. How an organism makes and breaks down the major classes of organic biomolecules—carbohydrates, lipids, proteins, and nucleic acids—determines not only how well it can create

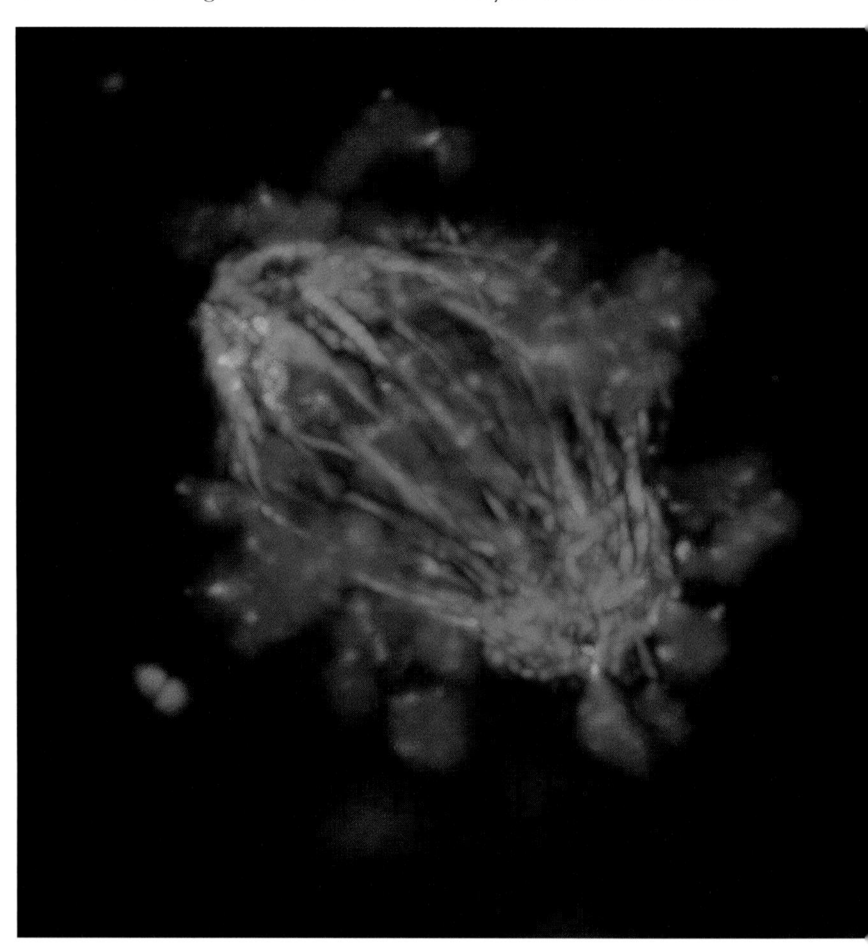

Above: An embryonic stem cell in mitosis. DNA is forming in this stem cell, a process that many scientists are eager to study, although the use of stem cells in research is currently hotly debated among politicians, doctors, and others. Top left: Single-celled organisms called bacteria are packed into dirt. They are crucial to the development of multicelled plants and creatures that derive sustenance from the earth.

and process energy but also how well it can reproduce.

At the cellular level, the sum of these processes can be called cellular metabolism. Again, all of these reactions eventually reach the cellular level, and individual cells need to be able to process nutrients to keep themselves and the larger organism alive. The human cell operates like its own little world. It has a membrane that is like a security fence that permits some things to come in while keeping other things out. It needs supplies in order to survive. It has structures that keep it viable.

FROM MOUTH TO MITOCHONDRIA

Energy is not much good if it is not in a form that is usable. Just as gasoline isn't of much use to a living organism as a form of energy, food molecules, when they are first ingested, aren't of much use to individual cells until they have been broken

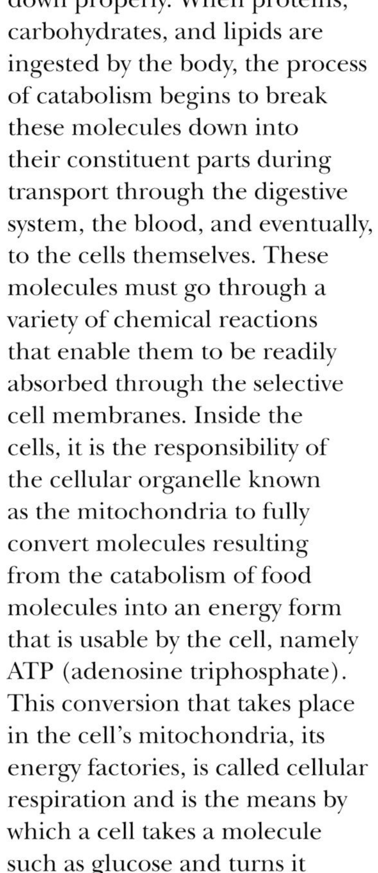

down properly. When proteins, carbohydrates, and lipids are ingested by the body, the process of catabolism begins to break these molecules down into their constituent parts during transport through the digestive system, the blood, and eventually, to the cells themselves. These molecules must go through a variety of chemical reactions that enable them to be readily absorbed through the selective cell membranes. Inside the cells, it is the responsibility of the cellular organelle known as the mitochondria to fully convert molecules resulting from the catabolism of food molecules into an energy form that is usable by the cell, namely ATP (adenosine triphosphate). This conversion that takes place in the cell's mitochondria, its energy factories, is called cellular respiration and is the means by which a cell takes a molecule such as glucose and turns it into a form of energy that the

This woman and her daughters share similarities on the cellular level, and as a result may have very similar metabolic rates.

cell can actually use. If this process takes place with oxygen, it is called aerobic respiration; without oxygen, the process is called anaerobic respiration. The means by which carbohydrates, lipids, and proteins are broken down within the body before and during the process of respiration are referred to as metabolic pathways. Glycolysis, for example, is the first step in aerobic respiration and is a metabolic pathway that breaks down glucose into a form that can be absorbed by the mitochondria.

RESEARCH AND BIOCHEMISTRY

The study of the chemical processes that take place at the cellular level in living organisms—especially how cells metabolize different molecules—is not done for empirical research alone. The applications of biochemistry are varied, ranging from the study of the effects of excess cholesterol in humans to AIDS research and treatment. By studying how bacterial cells metabolize penicillin, for example, biochemists were able to determine that the penicillin made it very difficult for the bacteria to build their cell walls.

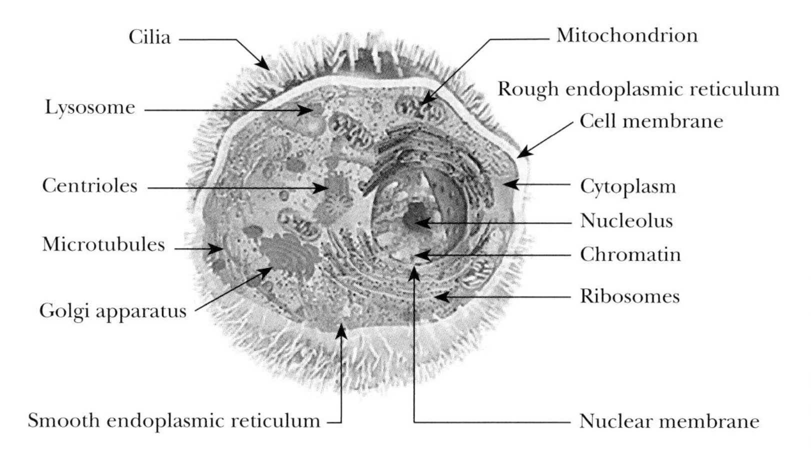

All cells have an outside membrane, a cytoplasm containing hundreds or thousands of organelles, and an inner nucleus. Each part of a cell functions differently toward an end product, like a microscopic factory.

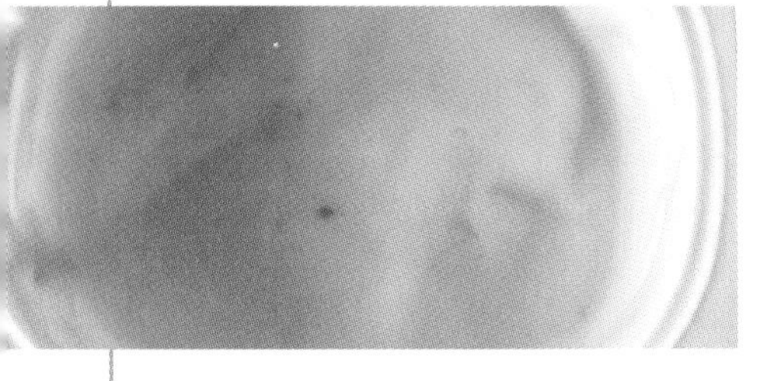

Enzymes

Earlier, we saw how RNA makes proteins that perform work. Well, here's a look at the agents of that work: enzymes. They do all the work in the cell.

An enzyme is a protein made of hundreds of amino acids that brings about desired chemical reactions in an organism. They work by dissolving chemical bonds between one molecule and another. For example, the enzyme amylase occurs in saliva and begins breaking down complex starches into smaller sugar molecules the second we pop a piece of bread or potato in our mouths. Thanks to enzymes, the digestion of food takes only hours instead of days, weeks, or years. Enzymes speed up reactions by factors of millions or even billions. The object that interacts with an enzyme is called a substrate; the altered substrate is referred to as a "product" of the enzyme. At first glance, enzymes sound a lot like catalysts, but they do differ. A catalyst is anything that brings about a chemical reaction; by this definition, heat can be a catalyst. An enzyme is a biologically derived macromolecular catalyst.

Of course, we have long exploited enzymes to work for us outside the human body. Whenever we bake bread, brew beer, ferment wine, or culture yogurt, we are using the enzymes of living organisms called yeasts to make these foodstuffs. Some cheese makers use an enzyme called rennet, taken from the stomachs of calves, to curdle milk. In the home, we use enzymes to break down food stains in our laundry. Or we may use packaged enzymes to tenderize meat before we cook it.

Above: This single-celled creature, called a Dendrocometes *protozoan, lives on the gills of freshwater fish. Biotechnology researchers hope to use creatures like these to produce enzymes. Top left: Yogurt is made using bacteria to ferment milk—this means that on a microscopic level, yogurt is alive!*

MECHANISM MODELS

In 1894 the German scientist and Nobel laureate Hermann Emil Fischer (1852–1919) tried to explain how enzymes work. He proposed the "lock-and-key" model of enzymes, which is that an enzyme has a highly specific shape cut into its side that perfectly fits its intended substrate. Like a key slipping snugly into a lock, the substrate fits into the enzyme, which then does its work. Then the new product disengages from the enzyme. Though this model was accepted for many years, modern scientists, observing these reactions

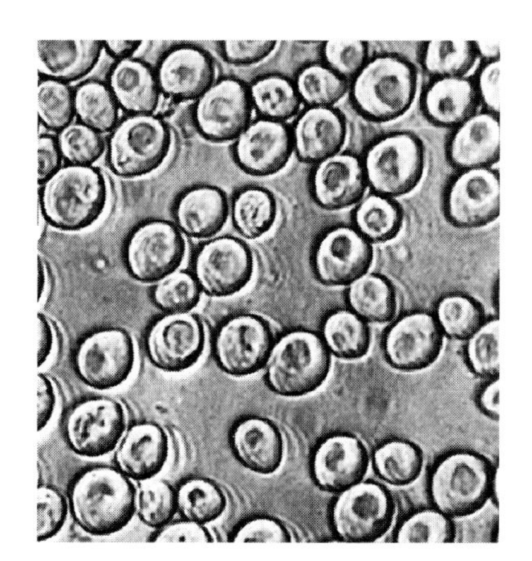
Enzymes, such as the yeast cells seen above, have been used in foods including bread, beer, and some types of cheese, for thousands of years.

in real time, have proposed the "induced-fit" model. In this model, the enzyme is not a rigid cookie-cutter shape. It's far more flexible. When the proper substrate sidles up to it, the enzyme can alter its shape to snugly envelop the substrate.

Regardless of the mechanism, the important thing is that enzymes will not act with just any substrate. Clearly, they are searching for their one true match, and will not trigger a reaction until they do. This choosiness plays an important role in metabolism. In a laboratory, a scientist mixes a catalyst into his test tube or beaker any time he wishes to achieve a reaction. Until the moment he starts his reaction, the two reactants are kept apart, separate and incapable of running amok.

In the human body, we do not have that luxury. Enzymes, in other words, are already in the test tube of the human cell. They are swimming around with a great number of substrates, and it would not do to have them initiating reactions with all of them. So nature has designed a mechanism—still something of a mystery to us—that keeps the enzyme's potential power under wraps until it finds a trigger that is chemically correct.

Some enzymes are easy to remember because their names are similar to the substrate they are breaking down. Maltose, a sugar, is made up of two glucose molecules held together by a single bond. Maltase, an enzyme, breaks that bond, leaving the two molecules free to travel on their own. The enzyme has taken a complex sugar and turned it into a less complex sugar, which is handy for digestion. In the same way, the milk sugar lactose is broken down by lactase. People who are lactose-intolerant do not have the proper enzyme, and thus cannot digest milk. Lipase, another enzyme, is responsible for breaking down triglycerides, complex fats or lipids that we consume in our diets.

At the heart of it, enzymes are nimble exemplars of how chemistry works in the body. If we can learn from enzymes, we may be able to develop drugs that more effectively destroy dangerous pathogens. Currently, a whole class of drugs, called inhibitors, work by blocking enzymes from carrying out their programmed function. They are helpful when enzymes are somehow interfering with a course of therapy.

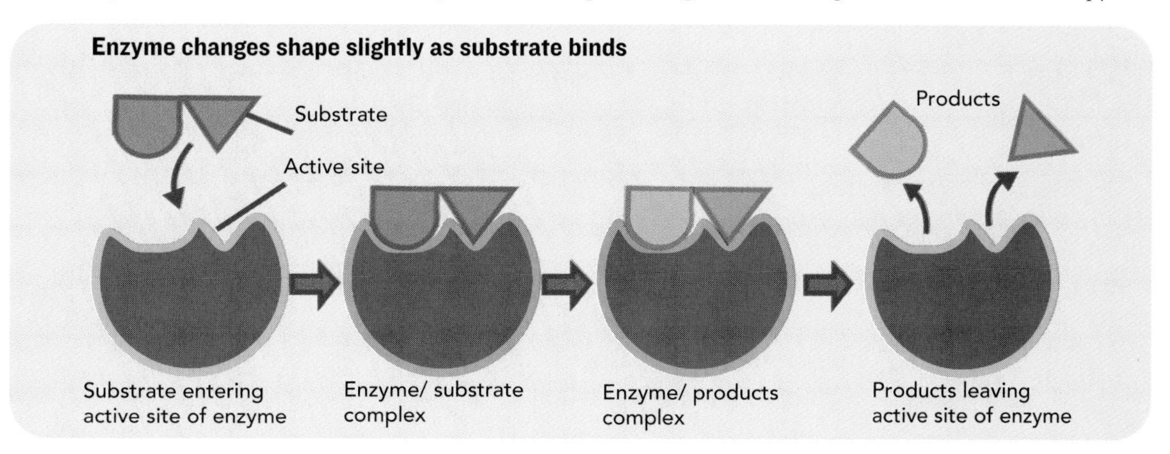

Substrates fit into enzymes like puzzle pieces, but are changed by the interaction into different forms. This process happens all the time, inside of our bodies and in countless other organisms, but we have only just begun to exploit it for medical and other purposes.

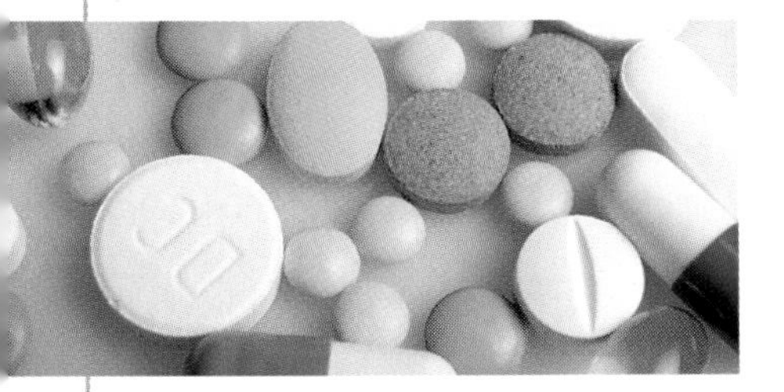

Medicine

The sight of an old willow tree hunched over a babbling brook, its weepy boughs dangling in the breeze, evokes a feeling of dreaminess more than cold, rational thought. Yet the willow tree is mixed up in one of the most interesting tales of modern medicine: the creation of lab-grown drugs and the birth of the pharmaceutical industry.

As far back as the fifth century BCE, the Greek physician Hippocrates recognized that an herbal remedy made from the bark of willow trees could ease pain and reduce fever. He was not alone. Other ancient peoples, such as the Sumerians, prescribed this medicine. Native Americans administered it to ailing tribesmen as well. For at least 1,800 years, humans had managed pain using this cure, but it wasn't until 1828 that the plant's healing chemical was isolated and named *salicin.*

From there a number of scientists in the nineteenth century experimented with the chemical's derivative, salicylic acid, which was used as an analgesic, though it had bad side effects. It upset people's stomachs, sometimes leading to diarrhea and bleeding of the stomach. An overdose of the medication could kill. Various chemists tackled the problem, trying to isolate the beneficial elements of salicin and eliminate the potentially damaging ones. Nothing seemed to work, until a chemist working in Germany, Felix Hoffmann (1868–1946), concocted a new version of the drug, acetylsalicylic acid, while searching for something to ease his father's arthritis. In a sense, he had improved upon nature because his compound was the first workable synthetic drug. It

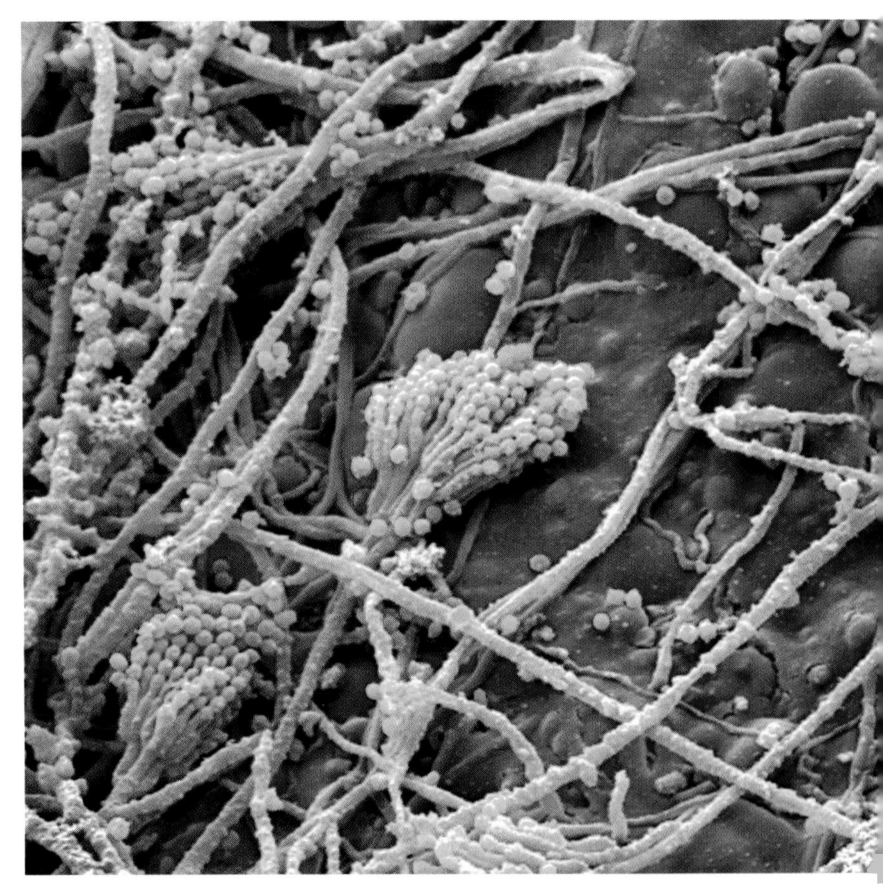

Above: Mold growing on bread is a common, if usually undesired, phenomenon. One of the most important recent medical developments, penicillin, owes its existence to an accidental discovery of the properties of mold. Top left: Medicinal pills are a recent development in a long history of curing disease with external supplements. Researchers continue to construct new and improved compounds of natural and synthetic materials, always pushing our knowledge of chemistry.

Willow trees like these have a surprisingly long and ubiquitous place in human medicine. Early societies from Greece to the Americas used willow bark to relieve pain, and it is a synthetic version of a chemical in willow bark that exists in many pain relievers today.

eased pain with little to no side effects. Delighted, Hoffmann's employer marketed the medication. His name was Friedrich Bayer (1825–80). The market name of the drug—if you haven't already guessed—was Aspirin.

Not until 1971 did scientists unlock the mechanism by which aspirin works. It is an inhibitor, or enzyme-blocker; that is, it prevents a key enzyme in the body from creating products called prostaglandins—the chemicals that send the pain response to our brains. They also cause inflammation, or swelling, of hurt or damaged tissues. When aspirin enters the bloodstream, it sticks to the enzyme cyclooxygenase and stops it from making more prostaglandin. The brain stops experiencing pain, the swelling goes down, and the patient feels a little better. Of course, aspirin works only for minor aches and pains, and is easily excreted from the body. To continue the beneficial effects, a patient must continue taking it if the illness or pain persists.

Aspirin's journey from folk remedy to over-the-counter medication exemplifies the common rite of passage in which science transforms naturally occurring chemicals into substances that can be relied upon to provide effective relief. Regardless of the drug or therapy, there is always a period of trial and error. Modern science requires drugs to undergo strict testing before they are released to the public. Typically, a new drug is designed in theory, modeled on computers, studied in test tubes, then tested on rodents, followed by larger mammals such as pigs or dogs, and finally on humans. The human trials alone may take several years, as researchers gauge the medication's effect on an increasingly larger pool of people to ensure safety and efficacy.

Historically, science has taken its cues from nature. Penicillin, one of the great medical discoveries of all time, occurred when Scotsman Alexander Fleming (1881–1955) observed the behavior of mold that had contaminated his lab specimen of the bacteria *Staphylococcus* actually dissolved the bacteria. Fleming knew well that living organisms developed self-protecting mechanisms against disease. Seven years earlier, he had discovered that lyzozyme, an enzyme in tears, routinely destroyed the cell walls of bacteria, protecting the human eye from infection. Modern medical researchers continue in this tradition, finding ways to augment and amplify the body's already impressive arsenal of protective chemicals and enzymes.

Agricultural Chemicals

The agricultural industry has long looked for ways to maximize crop production while at the same time minimizing crop loss. Many large-scale farmers use chemicals to achieve these goals, and this has created a booming industry for pesticides, herbicides, and fertilizers. These chemical advances are also appealing to individuals who desire a velvety green carpet of grass surrounding their home, and gardens full of beautiful flowers in beds permanently rid of weeds and bugs. Pesticides, herbicides, and fertilizers work on the chemical processes of animals and plants—their metabolism, for example—to prevent or encourage survival.

PESTICIDES

Pesticides work on unwanted pests in a variety of ways, from dissolving the protective coverings on insects or blocking their respiratory functions to altering their metabolism or nervous system. Organophosphorous pesticides are a staple of the pesticide industry and became widely used in the 1980s. They are used in both commercial and residential settings and work by interfering with the normal functioning of an insect's nervous system, namely by compromising a specific enzyme's ability to function. This enzyme, cholinesterase, breaks down acetylcholine in the nervous system and works as a regulator for nerve impulses. Acetylcholine is a neurotransmitter that carries information between nerves and between nerves and muscles. And without properly functioning cholinesterase, acetylcholine is not broken down, preventing nerves from resetting once a signal has been transmitted. There is no regulation of the firing of this information between nerve cells. The body can go into overdrive—twitching, convulsing—and this can result in death. Though deadly to small creatures, organophosphates in small doses are not deemed lethal in humans. But overexposure can cause detrimental effects on humans, too.

Above: The spraying of pesticides and herbicides, while enabling greater successes for agriculturalists, is frequently detrimental to humans, animals, and the general ecosystem, and is thus being constantly reevaluated. Top left: Wheat is a major agricultural product of several countries and is one of the crops frequently inundated with protective chemicals.

HERBICIDES

Herbicides also have a variety of ways in which they work to inhibit the growth of the plants they target. A large percentage of herbicides in use both commercially and by homeowners are based on glyphosphate. And, similar to pesticides, these herbicides work on the enzyme activity of their targets. Glyphosphate interferes with EPSP synthase, which enables the plants to produce necessary proteins. Other ways herbicides attack plants can include disrupting the plant's ability to photosynthesize or blocking the possibility of seed germination or the release of spores.

FERTILIZERS

In addition to hydrogen, carbon, and oxygen, plants need other nutrients in order to thrive. Three important ones are nitrogen, phosphorous, and potassium—without these, plants simply cannot grow normally. The three numbers visible on bags of fertilizer sold at any local garden store represent the percentages of the nitrogen, phosphorous, and potassium (respectively) present in the fertilizer mix. By knowing the existing soil chemistry before purchasing fertilizer, gardeners can make better, more informed choices. These nutrients also come from decaying plant matter, which is why mulching is a popular, effective, cost-saving, and sustainable means of fertilizing plants.

FINDING A BALANCE

As is the case with many scientific advances, there are both positives and negatives associated with the use of various pesticides and herbicides. DDT, for example, considered the first of the modern pesticides, is largely responsible for the eradication of malaria in the United States. However, the downside of DDT eventually became apparent, and its harmful effects—on birds and other living things—were brought to light by Rachel Carson's groundbreaking book *Silent Spring*. The chemical was banned by the EPA in 1972 and is no longer in use in the United States.

As a result of this and other lessons, the agricultural chemical industry continually reevaluates products for potential harmful effects and weighs these effects against any benefits. New limits and restrictions are routinely being considered and put into effects and alternatives are always being investigated. The use of chlorpyrifos to kill termites and the residential use of diazinon, an insecticide, were phased out just within the last several years. The overuse of permitted pesticides can also damage other living organisms—both plants and animals—that are not the targets of the treatment. Over time, this can affect the biodiversity of an area. And runoff from excessive use of sanctioned pesticides, herbicides, and fertilizers can end up in the soil, our water supply, and the air. The responsibility, therefore, of individuals to use products in the recommended doses is also a must in order to keep the risks from outweighing the benefits.

Birds, like this robin, are frequently the unintended victims of pesticides and herbicides, as they ingest the chemicals coating their food.

Termites, which can cause severe damage to houses, are often the targets of attempted extermination.

Unwanted Effects

A professionally cleaned carpet. A smoky bar. An award-winning rose garden. A spotless oven. A patio free of wasps and hornets. All of these serve as an example of the countless ways in which we come into contact with chemicals on a daily basis. As living creatures, we are dependent upon and strongly affected by changes in our environment. These changes may be obvious and immediately evident, or they may be more subtle, requiring time to build up in our surroundings before we become aware of the effect they are having on us.

TRICKLE DOWN, ALL AROUND

In addition to the harmful chemicals we come in contact with directly every day, other toxins easily travel into the sewage system through improper disposal, and into water and air supplies via runoff. Pesticides and fertilizers often end up in the soil and water supply. A variety of studies over the years have revealed the presence of pesticides and other toxins from household chemicals in urine and breast milk. Children, due to their lower body weight and growing, changing bodies, are particularly susceptible to the dangers presented by these chemicals. There is a great deal of concern that increased rates of diseases like cancer in children and the increased risk of breast cancer in women in the last 50 years are related to the environment in which we all live. Organic compounds can remain in the environment and be gradually absorbed over time. "Persistent organic pollutants" is a phrase used to describe organic compounds that result from various industrial uses that persist in the environment, often finding their way into the fatty tissue of living beings.

Above: Improperly disposed of containers such as these barrels can pollute entire ecosystems. Top left: The result of pollutants can often be fatal, not only to humans but to the many other living beings that share the planet.

MORE CHOICES

Chemical research is providing greater options than ever for consumers who want to do what they can to limit their exposure to those chemicals that may be more toxic. This is not just limited to increased availability of natural household products. An excellent example of how extensive chemical research has made it easier for humans to avoid dangerous toxins can be found in increased smoking legislation. When studies performed by scientists—among them biochemists—revealed that nonsmokers exposed to smoke can suffer some of the same detrimental effects as those with a habit, it was enough to set legislative wheels in motion to ban smoking in many public places.

WHAT INDIVIDUALS CAN DO

There is a lot that we, living in modern society, do not have control over. But beyond the purchasing of different, less toxic products, consumers can get into some good habits that might help reduce the toxins they bring into their lives. Many of us come in contact with chemicals through cleaning and gardening supplies. Individuals need to take care when applying "safe" fertilizers and using pesticides at home. The disposal of any toxic chemical needs to be handled properly, so that it does not end up in the water and soil. There are also steps that all consumers can take to reduce the need for using these additional chemicals at home in the first place. Lawn clippings, for example, rather than being bagged and added to a landfill, make a wonderful green, natural mulch for the other plants in the yard and can eliminate or reduce the need for chemical fertilizers. Choosing native plants—plants that will naturally thrive in a particular area—also means that the need for fertilizers will be greatly reduced. Some plants are natural deterrents to pests. Citronella plants, for example, help keep mosquitoes at bay. If pesticides and fertilizers are used, they should be used only in the appropriate amounts. A simple soil test, for example, can be affordably purchased at almost any lawn or home store and will ensure that excess fertilizer is not being used. And when fertilizers and pesticides are deemed necessary, they should never be used when rain is predicted. The disposal of all household chemicals, including personal medicines, should always be done with great care. The flushing of prescription medicine that is no longer needed is unfortunately a common practice and releases immeasurable chemicals into our water supply. Individual responsibility cannot be underestimated in efforts to make our environment more toxin-free.

Water polluted with human-produced chemicals has probably killed these salmon and may well claim other water-dependant life forms, including other fish and waterfowl.

The use of natural fertilizers and composting in gardens and fields is one way humans can reduce the use and spread of potentially harmful chemicals.

CHAPTER 9

CHEMISTRY SURROUNDS US

Left: We are able to blow soap bubbles because when water and soap are combined, the molecular composition of the resulting mixture is flexible enough to stretch into a bubble. Top: The aroma of frying potatoes can be a natural by-product of the cooking process—or a result of chemicals added to the mix by fast-food restaurants. Bottom: Even the paints we use inside and outside of our homes are the products of chemical reactions.

To take for granted the chemical reactions that pervade every moment of every day of our lives is to take for granted the food that we cook, the soaps we use to clean our clothes and our bodies, the paints that grace our walls, and the lights that illuminate them. A quick tour around an average home will show chemistry at work in everyday life: An antacid tablet relieves an upset stomach by neutralizing overactive acids in the esophagus. The heat-activated conversion of proteins and sugars into a caramelized crust helps sear meat for dinner and releases aromas that beckon families to the table.

Chemistry is also used commercially, sometimes for nefarious purposes. Some fast foods are packed with chemical compounds that give them the appearance, scent, and even taste of home-cooked meals. This is all thanks to an army of flavor and fragrance scientists that concoct tastes and smells that can fool the most heightened of human senses.

Beyond the innumerable chemical processes taking place within our bodies, it would be difficult to get through a single day without the assistance of chemical reactions that simplify and enhance the most mundane of tasks.

Chemistry in the Kitchen

When a chef walks into a kitchen, he or she may as well be stepping inside a chemistry lab. The ingredients of any recipe—salt, sugar, oils, fruits and vegetables, and meats—may seem like common foodstuffs, but at the molecular level they are composed of the very same natural chemicals we have seen throughout this book. In order to survive, humans need essential nutrients, such as carbohydrates, fats and oils, and proteins. Sugar, for example, is a carbohydrate, but a chemist would recognize it as sucrose, a chemical compound made up of carbon, oxygen, and hydrogen. Salt, a mineral nutrient essential for life, is sodium chloride, tiny crystalline cubes of the elements sodium and chlorine, held together by ionic bonds.

In recent years, a new field called molecular gastronomy has sprung up to better analyze the chemical properties of food. Molecular gastronomists

Above: The process of leavening bread happens when yeast—a living organism—feeds on a sugar and releases carbon dioxide as a waste product. The carbon dioxide is trapped in the batter as bubbles and causes the bread dough to rise. Top left: Sucrose, glucose, and fructose are the chemical names of the sugars found in fruits and vegetables.

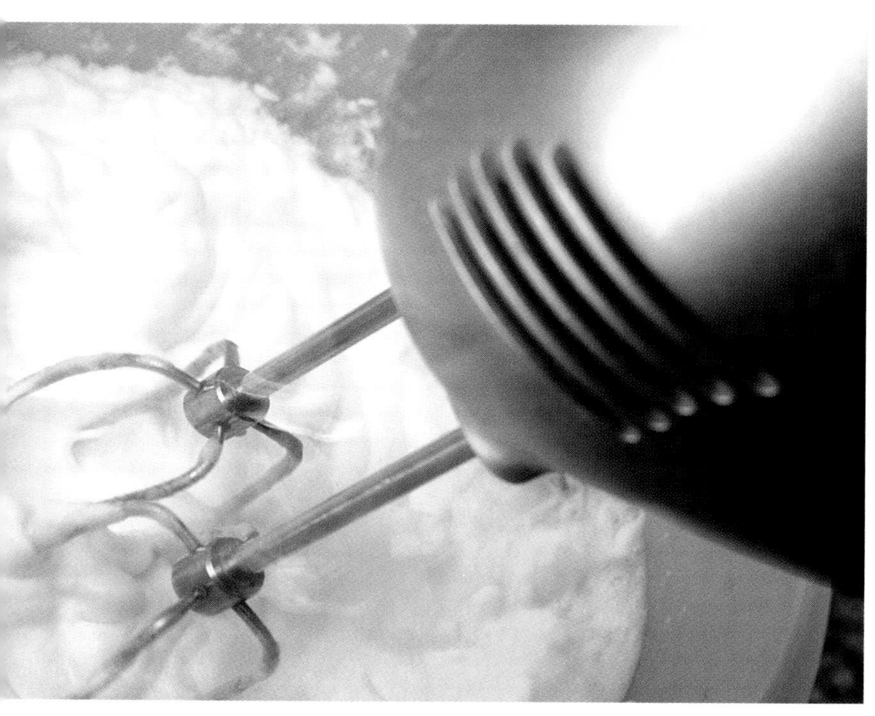

Egg whites become fluffy and then stiff when they are beaten for an extended period. As air is whipped into the eggs, amino acid chains in the whites break down, trapping increasing amounts of air and growing larger—and fluffier.

are often chefs with scientific training. Of course, one doesn't need to be a scientist to prepare a good meal. When a cook makes a delicious meal for friends and family, he or she is calling upon the natural chemical properties of various foods, mixing them in such a way that they interact and create new flavors or behave in desirable new ways. Thankfully, the chemicals in food behave the same way every time, ensuring a reasonable degree of success.

Let's look at some of these well-known reactions. To make a loaf of bread, a baker adds yeast—tiny, living cells—into the mix of flour, warm water, salt, and other flavorings. Activated by the temperature of the water, the yeast cells are awakened from their slumber and feast on simple sugars in the flour, producing carbon dioxide in the process. This gas, the very same one we expel when we exhale, creates tiny air pockets in the wet dough, puffing it up and making it rise. Without yeast and its gassy by-product, we would find it very difficult to bake light, fluffy bread.

When a cook breaks an egg, the white and yolk are runny, slipping and sloshing around in the bowl. The whites are made up of proteins in the shape of long chains of amino acids. When the cook beats the whites, he is mixing oxygen into the egg proteins.

Many fast food chains use synthetic compounds to generate specific smells and flavors in foods including hamburgers and french fries.

CHEMISTRY IN THE FAST FOOD LANE

If fast food is so unhealthy, how can it possibly smell so good? The savory scents that emanate from chain restaurants seem capable of wafting for miles, carrying an almost addictive aroma to passersby. The power of the aroma of any food is not to be underestimated. How food tastes is believed to be more directly related to the way that it smells rather than the interaction the food has with taste buds, because our olfactory systems are capable of recognizing thousands more different smells than our taste buds are able to discern tastes. With this in mind, the chemical flavor industry tackles both senses seriously, imparting smells and tastes—smoky, sweet, sautéed—that are mixed in laboratories long before they land in a kitchen. A growing number of companies create and combine chemical compounds that generate specific smells and flavors in foods, from strawberries to soda pop, from french fries to frankfurters.

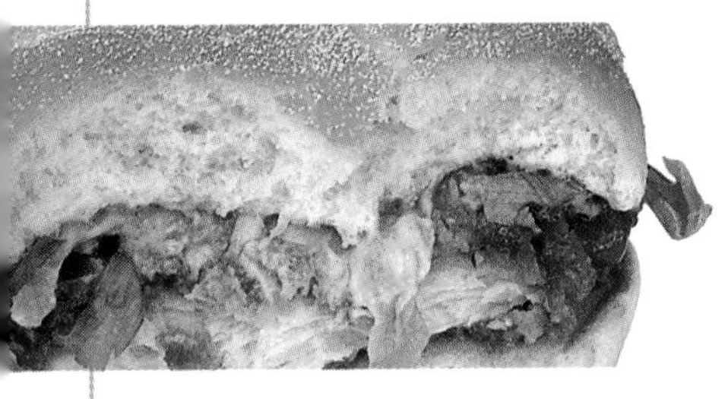

Chemistry in the Kitchen (continued)

As more and more air becomes incorporated into the proteins, the amino acid chains unravel. They become more and more disorganized, trapping air, increasing in size, and getting increasingly stiffer. Before long, the whites of a single egg can increase eightfold and become so stiff that the chef can turn his bowl upside down and the freshly whipped whites, now called meringue if sugar has been added, will not fall out.

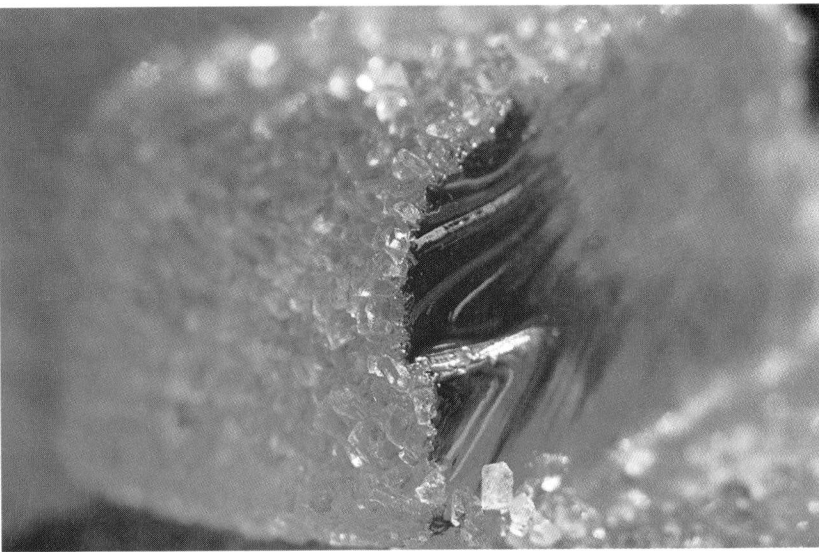

Sugar is a star player in kitchen chemistry. It gives candy its sweet flavor (top) and plays a part in a process called the Maillard reaction. When meat is heated (bottom), its proteins break down and combine with sugars in the fat to create many new flavor compounds. Top left: All food preparation is chemistry. Whether it is frying a chicken breast or boiling pasta, cooking causes food to undergo chemical changes.

CARAMELIZATION AND THE MAILLARD REACTION

In the wild, predators eat their kills raw. But humans prefer to cook meat. Why is that? It's because we have stumbled across a fascinating truth: Browned meat tastes delicious! When heat is applied to meat, the proteins on the surface break down and combine with the sugars present in the fats. The meat browns, the air is perfumed with a "meaty" odor, and a new flavor emerges that does not exist in raw meat. This process is called the Maillard reaction, named after the French physician and chemist Louise-Camille Maillard (1878–1936), who discovered that amino acids and sugar produce an appealing brown flavor when heated. This chemical reaction—also present in toast and beer—has since been exploited in the commercial food industry to produce artificially flavored products, such as sugar-based "maple" syrup. A similar browning reaction, called caramelization, occurs when sugars are heated. The longer the sugars remain over heat, the darker they become and the stronger their flavor. This process is used in candy-making and to develop flavor in vegetables, which contain natural sugars.

Fermentation was one of the first chemical processes harnessed by civilization. It results when yeast consumes simple sugars and then gives off carbon dioxide and ethanol (alcohol).

Water begins to boil at temperatures lower than 212°F because the atmosphere exerts less pressure on water at elevations higher than sea level. For instance, at 5,000 feet, water comes to a boil at 203°F.

WHEN A THREE-MINUTE EGG ISN'T

When does a three-minute egg take three-and-a-half minutes to cook properly? When it is being cooked at high altitudes. Cooks who have traveled or lived at high altitudes may be surprised at how quickly a pot of water comes to a boil but that cooking three-minute eggs for three minutes does not always do the trick. This is because even though the water is boiling, it is not as hot as water boiling at lower altitudes. The atmospheric pressure exerted on water is part of what determines how much energy is needed for it to boil. When you are cooking, this energy is in the form of a heat source beneath your pot. The higher the pressure, the harder it is for molecules to break the bonds that bind them together and break free from the surface of the liquid. The higher the altitude, the less atmospheric pressure is exerted on the surface of the water. This means that less heat is needed to cause the individual water molecules to break free of one another and escape the water's surface. The temperature at which water boils can differ almost 10 degrees Fahrenheit depending on whether it is boiling on a stove in Charleston, South Carolina, or Denver, Colorado.

Color and Light

High above the floor of the Sistine Chapel in Vatican City is the famous ceiling painted by the great artist Michelangelo Buonarroti beginning in 1508. Though they are 500 years old, the pigments in Michelangelo's scenes are as bright as the day he applied them. Artists of his day did not have access to store-bought paints; they made their own using various natural substances such as soft rock, vegetable matter, and the bodies of insects and other animals. These pigments work much the same way as modern pigments do. In essence, they absorb all light except a specific, desired wavelength, such as brown, blue, or red. When early artists discovered they could mix pigments with substances to help them spread ("diluents") and to help them stick to surfaces ("binders"), a world of decorative and artistic painting was born.

As understanding of chemistry advanced, people were able to devise chemical formulas to create synthetic pigments needed in modern life. In 1856, a young Englishman named William Henry Perkin (1838–1907) developed the first dye made from an organic chemical compound, called aniline. Perkin, who was only 18 years old, called his rich purple dye "mauveine." We know the color as mauve. Today's chemical compounds form the basis of innumerable paints, which often can be mixed fairly quickly once the desired color is analyzed. Paint swatches are sometimes mixed according to the Munsell color system, which arranges a variety of colors along a color wheel. The colors in the book you are holding are actually a mix of four different chemical-based inks used by printers: magenta (light red), cyan (blue), yellow, and black. These four colors can be mixed to make all others on the printed page.

LIGHT-EMITTING DIODES

It is impossible to divorce color from light, since all color is derived from the light spectrum. If you have ever touched an illuminated lightbulb, you more than likely pulled your hand away for fear of scorching your fingers. Incandescents are not

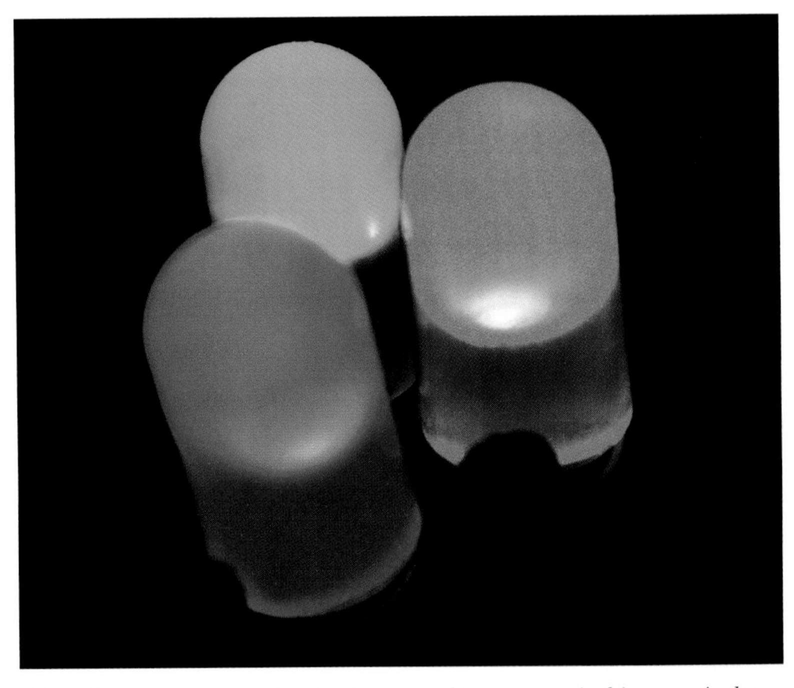

Above: When the atoms and molecules in an object are energized in a particular way, the result is light—and heat. Some light sources, light-emitting diodes (LEDs) for instance, give off less heat and are therefore more efficient. Top left: Paint chemists work to identify and create pigments to produce specific colors.

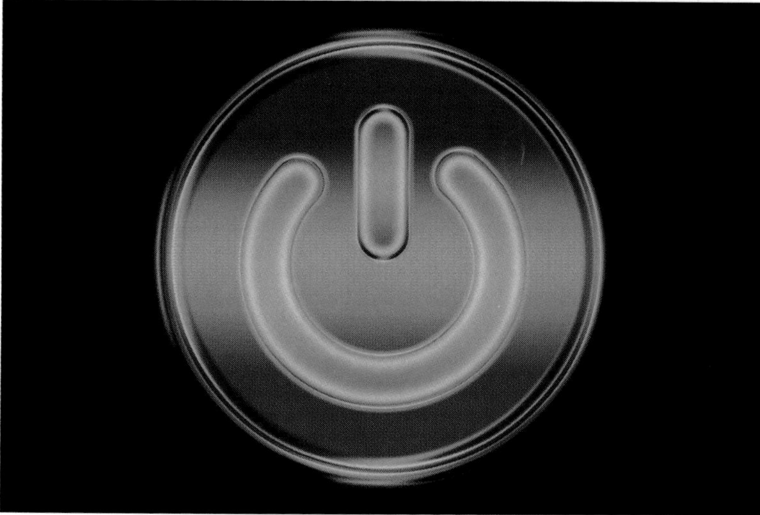

Light can be produced by a number of chemical reactions. Neon light (top) is produced by sending an electrical current through a gas. LEDs (above) work by electricity being passed through a gallium arsenide chip, which glows and produces light.

very efficient. Yes, they produce white light, but much of the energy they consume is used to crank out excess heat, which simply wafts off the bulbs and is wasted. Fluorescent bulbs are slightly more efficient, but even they get hot.

Light-emitting diodes (also known as LEDs) are probably the most efficient form of lighting used today. We know them as tiny "on-off" indicator lights found on computers, answering machines, televisions, and a host of other electronic devices. The brain of such a bulb is a piece of electronic circuitry fashioned out of a chemical compound called gallium arsenide. When electric current is applied to the two prongs sticking out of the bulb, the electronic "chip" produces infrared (invisible) light or colored (visible) light. Infrared LEDs are used to send remote-control signals to devices such as television sets. Colored LEDs, meanwhile, are used to tell us when a particular appliance is switched on.

Scientists dream of creating an extremely powerful LED bulb that will emit white light without heat and thus be suitable for lighting homes. Bulbs like this are already in the works. Such LEDs will be smaller than a dime and very economical: A two-inch-wide electronic chip can be sliced up into 10,000 separate lights. Since they are 10 times more efficient than an Edison bulb, and can last 100,000 hours or 12 years, they could reduce America's total electricity consumption by 10 percent a year, which means we wouldn't need to build a power plant for another 15 years.

PLAYFUL NEON

Other chemicals produce more familiar lights. Gases such as neon, argon, and krypton are known as the noble gases and are found on the far right column of the periodic table (see Ready Reference). These gases have a remarkable number of applications in a host of industries. When pumped into narrow glass tubing and excited with electric current, each gas produces a distinctive, brilliant glow. Neon, for instance, emits red light. This gas is responsible for the bright, sometimes garish signage that has become synonymous with commercial districts all over the world.

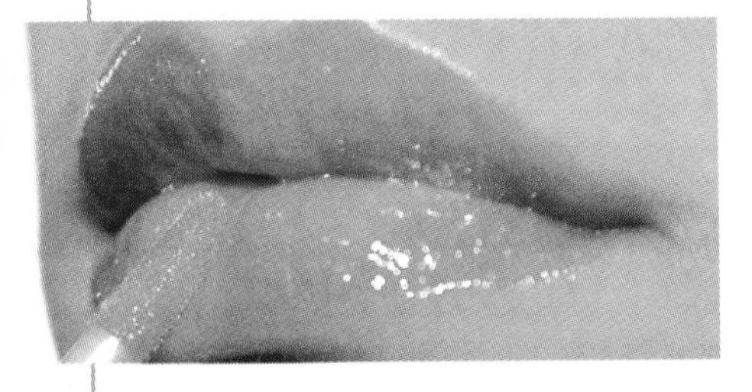

Cosmetics and Perfumes

Human beings have always been willing to pay a high price for beauty, employing the chemical reactions of naturally occurring elements to create colorful combinations that would adhere to the skin and hair. The ancient Greeks and Egyptians used, among other things, lead carbonate, or "white" lead, a naturally occurring powder, to lighten their complexions. However, their desire to be fair led to poisonous and even deadly complications when the lead was absorbed through their skin. Today, billions of dollars feed a cosmetics industry dependent on advancements in chemistry and biotechnology to continue to provide the allure of the enhanced appearance of skin, hair, and scent.

Above: Because of its chemical composition, beeswax is a common ingredient in cosmetics ranging from lipsticks to moisturizers. Top left: Creating cosmetics is more than simply choosing appealing colors; it is a complex chemical process.

LIP SERVICE

For many, the sheen of rosy red lips is not just sought after, but considered a necessity. Lipstick has to accomplish more than is usually taken into consideration. It must be smooth and easy to spread. The color must be able to be mixed into the lipstick but also needs to be insoluble in water, so that when it comes in contact with human saliva or a beverage it doesn't lose its lustrous color. Although the ingredients in lipstick vary, and can include everything from fish scales to gold dust, it is basically composed of a combination of wax, oils, and dyes or pigments. Beeswax, for example, can be combined with nonvolatile oil, such as castor oil, to create a substance that can be spread but will still keep its shape while in the lipstick tube. The pigments that are used may be naturally occurring, such as iron oxide—more commonly known as rust—or can be synthetically developed. The chemistry at work here is not just limited to the production of the lipstick itself. Lipstick dyes react with proteins on the skin. The extent of the effect depends on the dye that is used, but the results can be startling. A lipstick that appears pale or clear in the tube can result in a color that is drastically different once it comes in contact with the proteins on the surface of the skin.

SKIN DEEP

Many moisturizers subscribe to the old adage "oil and water don't mix" in treating dry skin. Some moisturizers work by a process called occlusion, in which the oil-based ingredients essentially create an impermeable barrier on the surface of the skin that reduces the possibility of water loss. Other classes of moisturizer depend more on ingredients that operate as humectants, which are substances that can absorb water from their surroundings. Popular humectants found in many cosmetics are alpha-hydroxy acids, which are derived from various fruits. Containing hydrophilic or "water-loving" molecular groupings, these acids can bond with water molecules and therefore attract and keep more water in the outer layer of the epidermis.

OF NOSES AND ROSES

Fragrances, too, have long been in demand and were once painstakingly extracted by hand from the blossoms and roots of favorite flowers before being mixed with oils and then bottled. Natural essences such as lavender or rose are still extracted, though there are a variety of extraction techniques available today. Solvent extraction, for example, uses organic solvents such as dimethyl ether to extract the aromatic compounds from the natural material. The components found in perfumes may also be created through organic synthesis before being combined to create just the right fragrance.

The olfactory response varies greatly from person to person. Odorant molecules in the air bind to receptors in the lining of the nose, and these different bits of information are interpreted in the olfactory bulb region of the brain, which then distinguishes them as different odors. Individuals with an exceptional ability to differentiate even the slightest differences in fragrances are much sought after in the perfume industry. These "noses," or *nez* as they are called in France, sit and sniff the day away in front of a perfume organ, which is essentially a pallet packed with fragrances. There they combine different essences, each with a specific chemical compound, to find a combination that is just right. In most cases, one particular scent is created using hundreds of ingredients or components. The way a particular perfume smells on its wearer is also a demonstration of chemistry at work. When fragrances are applied to the skin, the numerous chemical compounds in the perfume interact with the compounds of the skin. Because each individual's body chemistry is unlike anyone else's, the way a perfume smells will vary from person to person.

A scent wafting from a field of lavender is interpreted by our brains through complex chemical interactions.

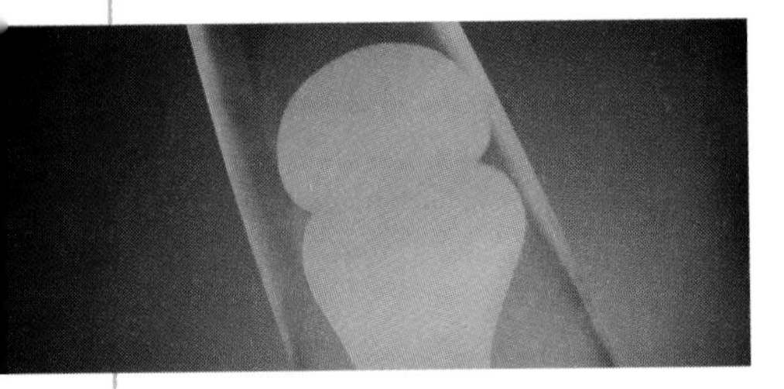

Mixing It Up

Birds of a feather flock together, as the saying goes. And in chemistry, as in many other relationships and interactions in life, some things go together and some just do not. In chemistry, this idea is expressed as "Like dissolves like." The ability of different chemical compounds to mix together is dependent on their similarity. Compounds with less in common tend not to want to be in close quarters. Those chemicals and compounds that do not want to mix on their own sometimes need to be helped along so that they can work together to achieve desired results, in the kitchen or in the laundry room.

Above: When two unblendable substances are mixed together it is called an emulsion. In A, the oil separates from the vinegar in a commercial salad dressing because they are dissimilar in chemical makeup. After they have been mixed, the oil disperses in the vinegar (B); however, they quickly separate again (C). Adding an emulsifying agent, such as soy lecithin, stabilizes the mixture so that the oil stays mixed longer (D). Top left: Lava lamps perfectly demonstrate that oil substances and water do not mix.

LIKE OIL AND WATER

It is commonly known that oil and water do not mix. The reason why lies in the differences between the molecular properties of the two substances, which direct the way they behave around other compounds. Water molecules are polar. This simply means that the overall charge of each water molecule is not evenly distributed, giving each "end" of the molecule a slightly negative or positive charge. So, the more positive "end" of a water molecule would be pulled toward a negative charge in its immediate vicinity, and likewise the more negative end of the molecule seeks out a more positive influence to pair up with. This polarity is not only what holds the water molecules to each other but is also the reason water reacts so well with a compound like table salt (sodium chloride or NaCl), which readily breaks down into positive sodium ions and negative chlorine ions, both of which are happily attracted to water's polar nature.

Oil, on the other hand, is nonpolar, and its charge is more or less evenly distributed among each of its component molecules. Without the same kind of positive-negative give and take that exists in water molecules, the two are immiscible. However, although oil and water may not mix, emulsifiers can help the two come to terms with their differences. In the kitchen, a common example of this is mayonnaise, in which oil and water form an emul-

A drop of water on a leaf is able to hold its shape because of surface tension—the tendency of the molecules in water to pull downward and toward the sides, creating as little surface area as possible.

sion with a little help from egg yolk, which acts as the emulsifier. Monoglyceride, which is formed by the reaction between fatty acids and glycerol, is a very common emulsifier and is an indispensable ingredient throughout the food industry.

CLEANING UP

The leggy hydrometra insect dances delicately across the surface of a lake. A droplet of water lands gently on a piece of fabric and maintains its shape—a smooth, intact, liquid sphere. Part of what makes both of these common occurrences possible is the surface tension of water. Surface tension is created by the interaction between the water molecules themselves. The polarity—unevenly distributed charge—within a water molecule allows for individual molecules to create bonds between one another. At the surface of the water, though, the molecules are no longer completely surrounded on all sides by their fellow polar waters. They have a tendency to pull down and to the side, in the direction of the other molecules and away from the surface.

The ability of soaps and detergents to clean well depends, in part, on how "wet" the water is, how well it spreads. Surfactants—compounds that "act" on the surface—help reduce the surface tension of water, allowing it to spread more evenly over the object being cleaned. Soap is a surfactant (essentially the same as an emulsifier, but surfactant is used in nonfood references) that aids in the emulsification of the water and grease. Soap molecules have both a hydrophilic and a hydrophobic end—one end is attracted toward the water molecules and pulls in that direction, while the other is more attracted to fats and oils. This binding of the soap molecules to both the water and the oil or grease allows for the soil and grime to be suspended and held long enough for the scrubbing action of hands and machines to do their work.

Above: Washing detergents and soaps are surfactants—they reduce the surface area of water, which in turn allows the water to be distributed more readily over an area being cleaned. In addition, they act as emulsifiers, helping the soap adhere to dirt and grease so they easily wash away.

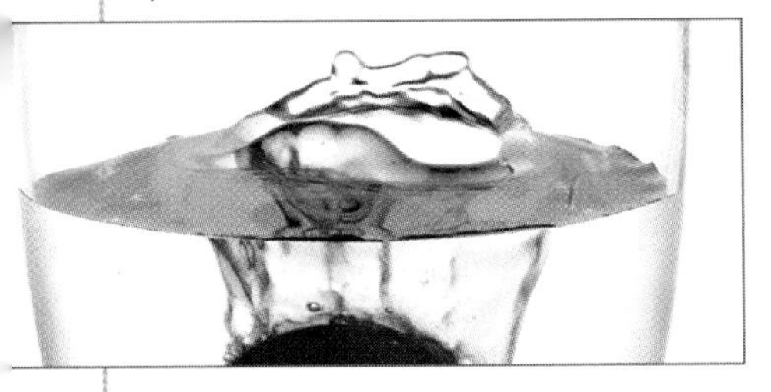

Acids, Bases, and Buffers

A few weeks into a harsh, blustery winter, every passerby seems to go around with chapped lips and skin rubbed raw by the constant assault of the elements. A baby's tender bottom can also develop a rash, and a cold sufferer may find his or her nose chafed by frequent wiping with tissues. Believe it or not, these common skin reactions can teach us much about the importance and power of acids and bases.

Human skin and hair are protected by a layer of sweat and lipids, or waxy fats, called sebum. This fine layer provides a kind of waterproof barrier against harmful bacteria that we encounter in our daily lives. This layer is referred to as the acid mantle because it is slightly acidic, with a pH value ranging from about 4 to 5.5. This is not a strong acidity, but it is just enough to destroy the bacteria that seek to enter our bodies on a daily basis. If the acid layer is disturbed in any way, the exposed skin dries, reddens, becomes rough, and flakes off. If the pH level rises, becoming more alkaline or basic, a person could suffer a possible skin infection.

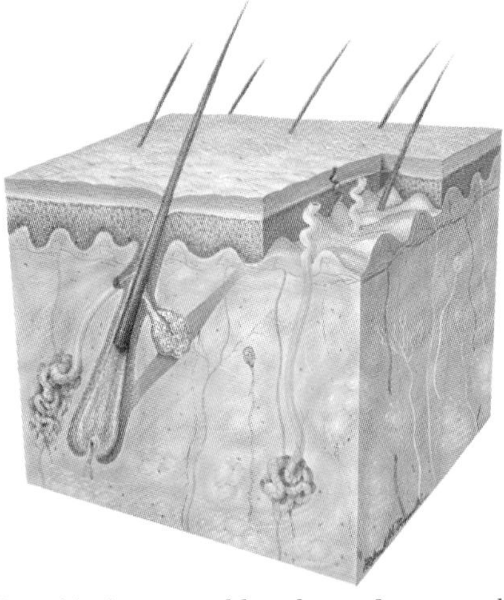

Our skin is protected by a layer of sweat and fats called the sebum. Ongoing chemical processes in our bodies help maintain a slight acidity on the skin's surface, which protects it from bacteria.

A person who has not bathed recently may complain of feeling "greasy," and regard a long, hot, sudsy shower as a good thing. But in fact, constant rubbing, frequent washing, or using alkaline soaps or detergents can all destroy the skin's natural acidity, leaving it vulnerable to disease and the elements. The same is true of human hair. Harsh soaps and shampoos can strip the acidic layer off each strand, leaving hair brittle and frizzy. This is why many personal care products are described as being "pH balanced"; if properly made, fine shampoos, soaps, and moisturizers can lower one's pH level and allow the body to restore its protective acid layer.

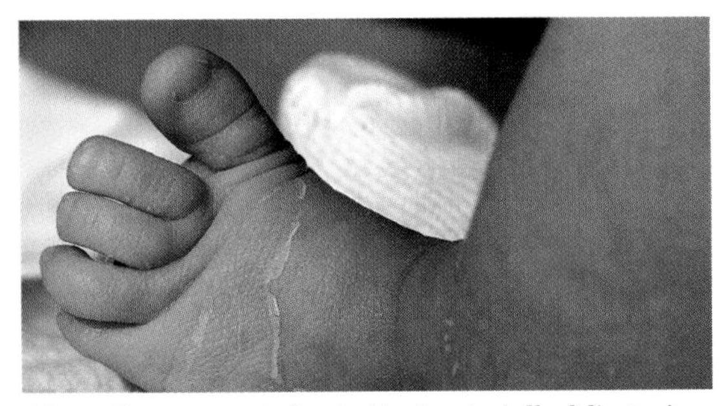

Above: Because an infant's skin is especially delicate, it requires gentle care to avoid washing or scrubbing away the sebum. Otherwise, dry, flaky skin can result. Top left: An antacid tablet, which is made up of citric acid and sodium bicarbonate, relieves heartburn by neutralizing excessive stomach acids.

THE UPS AND DOWNS OF ACIDS AND BASES

Since the human body is dynamic, or constantly changing, the levels of acids and bases rise and fall. But in order to function well, the body has developed tricks for keeping the correct pH at all times. For example, the interior of the stomach must always remain acidic in order to digest food. The intestines must always remain somewhat basic to counteract the effects of the stomach's contents as they pass by. To stay within the correct pH zone, the stomach and the intestines use buffers—special solutions that stop massive changes in pH levels.

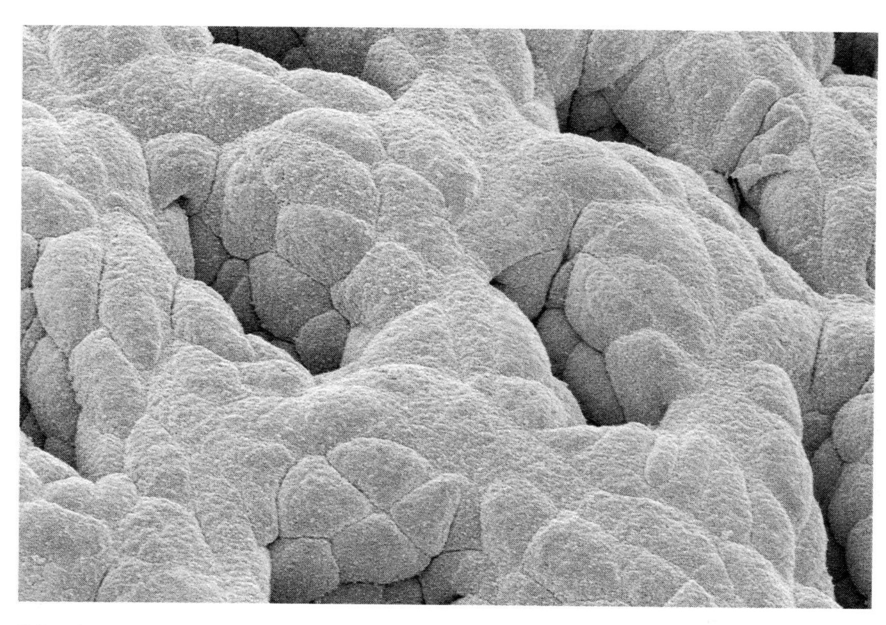

Digestion requires a host of chemical processes and reactions. The glandular lining of the stomach, shown here, secretes digestive chemicals that are highly acidic. Farther down the digestive tract, the intestines maintain an alkaline environment.

For example, our blood contains large amounts of carbonic acid (a weak acid) and bicarbonate (a base), which work together to make sure the blood is always at pH 7.4. If blood pH falls below 6.8 or rises above 7.8, one can become very sick or even die. The weak acid-base duo guard against this. In a pinch, the bicarbonate can neutralize—or render ineffectual—excess acids in the blood, while the carbonic acid neutralizes excess bases.

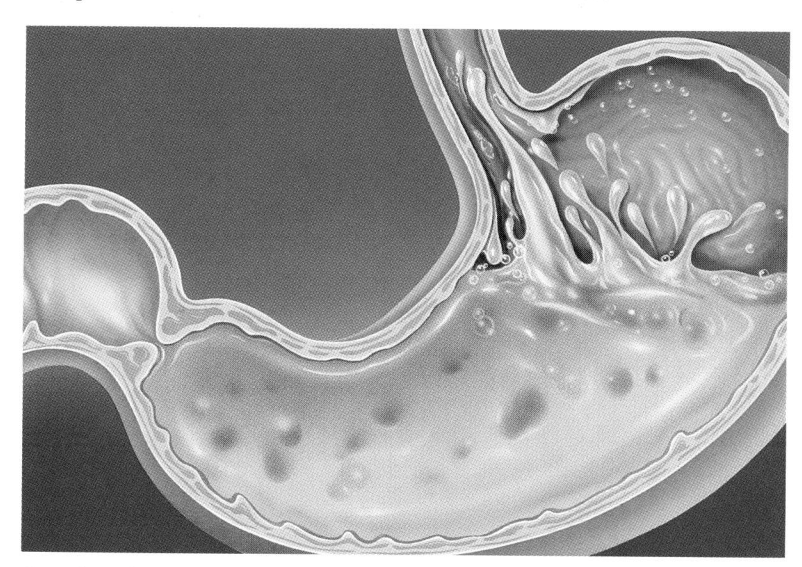

Sometimes the stomach secretes so much acid that some of it splashes into the lower part of the esophagus, causing heartburn. Antacids are the most common remedy for this condition. Made up of an acid and a base, antacids create a chemical reaction that relieves the discomfort.

In a sense, a similar process takes place when we ingest antacids or milk of magnesia. After a meal of rich foods, the stomach is stimulated to produce copious amounts of gastric acid, some of which may splash up into the lower end of the esophagus, causing an uncomfortable burning sensation. The sufferer may reach for an antacid tablet, which is made up of citric acid and sodium bicarbonate—an acid and a base. When the tablet dissolves in water, the two chemicals react by fizzing. Then, when consumed, the bases buffer the excess gastric acids by binding to them. This lessens the acids' painful effect on the body. The medication is an artificial way of mimicking the body's method of always bringing itself back into equilibrium.

CHAPTER 10

CHEMISTRY OF THE ENVIRONMENT

No matter what the environment or climate—tropical (left), arid (top), or polar (above)—chemistry plays an important role.

The systems that govern the stability of the planet and guide the interactions of the oceans and seas with the air above, and with the land below, all rely on chemical reactions. But there is a delicate balance to be maintained.

These systems are greatly affected by the organisms inhabiting the planet. Interaction with humans alone greatly impacts the chemistry that keeps these systems in balance. The study and protection of the environment can unite people from different cultures and regions. But it can also cause tension, as more technologically advanced countries often contribute more to environmental problems that affect individuals beyond their borders. New technology often presents as many problems as it does solve them. For example, the rash of new electronic gadgets available has contributed an increasing amount of toxic trash that is inundating our landfills. Chemists constantly work to understand the effects of such hazards and to develop means of mitigating their negative impact. The solution to many of the planet's ills lies in our ability to understand the chemical principles that govern the Earth and to work in harmony with them to find resolutions for these problems.

Carbon and Nitrogen Cycles

Above: Rain moves carbon through the biosphere. Excess carbon dioxide in the atmosphere has led to a gradual acidifying of the oceans as dissolving carbon dioxide increases the percentage of hydrogen ions in seawater. Top left: Trees absorb solar energy and carbon dioxide and later release carbon when ingested by animals or in the process of decaying. This taking up and releasing of carbon makes trees an integral part of the carbon cycle.

Our biosphere, those parts of our planet and our atmosphere that have the ability to support living organisms, is in a constant state of flux. The air we breathe is composed of many different chemical compounds, as is the soil beneath our feet and the water in our oceans and streams. Many of these compounds are absolutely necessary for the survival of our species and the other living organisms with which we share the planet. Other compounds, while crucial in a limited capacity, can become dangerous, even life threatening, when they exist in excess amounts. The balance and cyclical exchange of these compounds are vital to all ecosystems, the ecological communities of animals, plants, and living organisms that share the same environment.

AROUND WE GO

Clearly defined cycles describe how our most important chemical compounds make the journey from atmosphere to organism to water and soil and then eventually back to the atmosphere. Four cycles of significant importance to humans are the carbon cycle, the nitrogen cycle, the oxygen cycle, and the water cycle.

THE CARBON CYCLE

With every breath, we expel carbon dioxide from our lungs. Plants use carbon dioxide during the process of photosynthesis to create energy supplies for themselves and later transfer some of this energy to the animals that ingest them. Industry also contributes a great deal of carbon dioxide to the atmosphere, with a good percentage of it causing significant problems for our atmosphere. The carbon cycle defines the way carbon moves between its four main reservoirs: the atmosphere, the oceans, the terrestrial biosphere (the land and freshwater), and the sediments (carbon "inside" the Earth, mainly fossil fuels).

Carbon dioxide in the atmosphere is absorbed by trees and plants that will use it for photosynthesis. They also store carbon. Plants, like humans, also perform respiration, which returns carbon dioxide to the atmosphere. Plant matter is either eaten by animals, or the plant itself dies and decays. Dead organisms—plants and animals—are absorbed by the Earth, where they may eventually be covered by years

Although nitrogen is a naturally occurring element, it is also emitted in large amounts by burning fossil fuels, resulting in atmospheric pollution.

of sediment and, given many millions of years, become fossil fuel. Carbon also cycles through the planet's waters in a manner similar to what takes place on dry land. The plants in the ocean also use carbon dioxide for photosynthesis and store carbon, and the animals in the ocean also consume these plants for their energy. Similarly, aquatic plants and animals respire, returning carbon dioxide back to the atmosphere. When these animals and plants die and decay, they contribute carbon to the ocean floor, and eventually they too will be covered by sediment. Constant combustion of fossil fuels causes a major strain to the carbon cycle and has resulted in a drastic increase in the amount of carbon dioxide in the atmosphere over the last 100 years. (For more about the carbon cycle see page 112).

THE NITROGEN CYCLE

Nitrogen is the most abundant element in our atmosphere. It is found in decaying organisms and the waste produced by animals, and is a key chemical component in human beings—found in amino acids and nucleic acids, among many other places. Furthermore, without nitrogen, plant matter would be unable to survive. All these sources contribute nitrogen to the soil, and once it is there, two different bacteria work with it. Nitrogen-fixing bacteria "fix" or convert nitrogen into nitrates, nitrites, and ammonia, which can then be used by plants and animals. When plants and animals die and decay, they return nitrogen to the soil.

A second kind of bacteria present in soil, denitrifying bacteria, convert nitrates back into gas, and it is returned to the atmosphere. Too much nitrogen in the soil and water—from fertilizers and other pollutants—can make plants grow much too quickly and beyond the limitations of their habitat, ultimately causing them to die. (For more about the nitrogen cycle see page 113).

Legumes such as these green string beans are nitrogen fixers. Bacteria on small root nodules convert atmospheric nitrogen (N_2), which cannot be used by plants, to ammonium (NH_4) or nitrate (NO_3) ions. These forms of nitrogen can then be used for growth.

Atmosphere, the Oceans, and the Water Cycle

From air to land and back again, water's forms are constantly changing. Ice caps, snow, lakes, rivers, and water vapor are just a few of the guises water takes on as it weaves its way throughout the biosphere. The majority of freshwater on our planet is frozen in ice caps and snowfields, which are literally fields and expanses of permanent snow cover, while saltwater covers the planet in the form of oceans.

THE WATER CYCLE

The hydrological cycle, or water cycle, describes how water is recycled between Earth and its atmosphere. Most of the water that exists now has been around for hundreds of millions of years and has seen a lot of changes. There are five major actions that help water on its never-ending journey: condensation, precipitation, infiltration, runoff, and evapotranspiration.

As water in the atmosphere condenses into clouds, mol-

Above: This diagram provides a simple illustration of the hydrological cycle, or water cycle. Water circulates from the oceans, into the atmosphere, onto land, and then back to the oceans. Top left: Water is the most plentiful substance on Earth, and so it is no surprise that it plays a major role in the environment's chemistry.

Changes in ocean temperature or chemistry, such as those caused by pollution, can devastate coral reefs.

ecules combine and the water changes from vapor into liquid or ice until it eventually falls as precipitation, in a form such as rain or snow. The water may land directly in an ocean, river, or lake, or it may enter the oceans by way of runoff, taking its time, cascading down the side of a mountain before entering a stream that feeds into a river that will ultimately head out to sea. Precipitation is also absorbed by the soil through a process called infiltration and can exist as groundwater. Water may also move into a body of water directly from soil that has become saturated. No matter where water has found a home—ocean, puddle, or soil—it can be evaporated back into the atmosphere under the right conditions. Plants that have absorbed water through their roots can also return it to the atmosphere by way of transpiration through the stomata, or openings, in their leaves. The evaporation of water from soil and plant transpiration is more specifically referred to as evapotranspiration. And, of course, the human body contributes water to the atmosphere through pores as sweat, which will also evaporate. (For more about the water cycle see page 113).

WORLD OF WATER, WORLD OF WEATHER

Ninety-seven percent of Earth's water is contained in the oceans—a tremendous force on this planet. The atmosphere, though we may not realize it, is also a colossal reservoir of water. The oceans' delicate and sometimes tempestuous relationship with the atmosphere is primarily responsible for our climate, affecting everything from the global temperature to the winds. The atmosphere comes in contact with the ocean, coaxing its waters into waves and influencing its currents. The ocean, in turn, regulates much of our atmosphere's temperature. Water has what is referred to as a high specific heat, meaning that it takes an enormous amount of heat to raise water just one degree Celsius. This helps to explain why the raising of the average temperature of the oceans by just one degree is of such critical importance to the health of our planet. The ocean's thermal capacity makes it a massive reservoir of heat and energy and enables it to serve as a temperature buffer for the entire planet.

WORKING TOGETHER

None of Earth's systems—nitrogen, carbon, water, climate, and so on—operates independently. The ocean, which covers 70 percent of the planet, is a major player on the systems stage. Just one example of how systems interact is illustrated by the increasingly worrisome relationship between carbon dioxide and the ocean. As levels of atmospheric carbon increase, the ocean is naturally helping absorb them. It is estimated that the ocean absorbs roughly one-third of the carbon dioxide created by human actions. But the amount of carbon dioxide created is constantly increasing. Over time this causes acidification: The pH levels of the surface waters of the ocean gradually drop. Life is very sensitive to even the slightest of changes, and this change in ocean chemistry can have devastating effects on marine life, including vital organisms such as coral.

Working in tandem, the oceans and our atmosphere are responsible for climate and weather, including hurricanes.

Greenhouse Gases and Ozone Layers

The Earth is wrapped in a protective blanket that separates it from a vast and frigid atmosphere. Composed of a variety of layers, the atmosphere is home to numerous gases and chemical compounds that interact and work together to perform a variety of important functions, including regulating the planet's temperature and protecting its living organisms from the potentially harmful rays that emanate from the Sun. Affecting the efficiency of the atmosphere are several human-related factors. In the media we hear news about smog, and understand that the brownish haze blocking the Sun is bad. We see on weather reports that ozone pollution is bad, too, but also know that the ozone layer is necessary and in danger. Then, of course, there are greenhouse gases and global warming. All of these must be taken into account together to understand their cumulative effects and how they interact to sustain life on Earth.

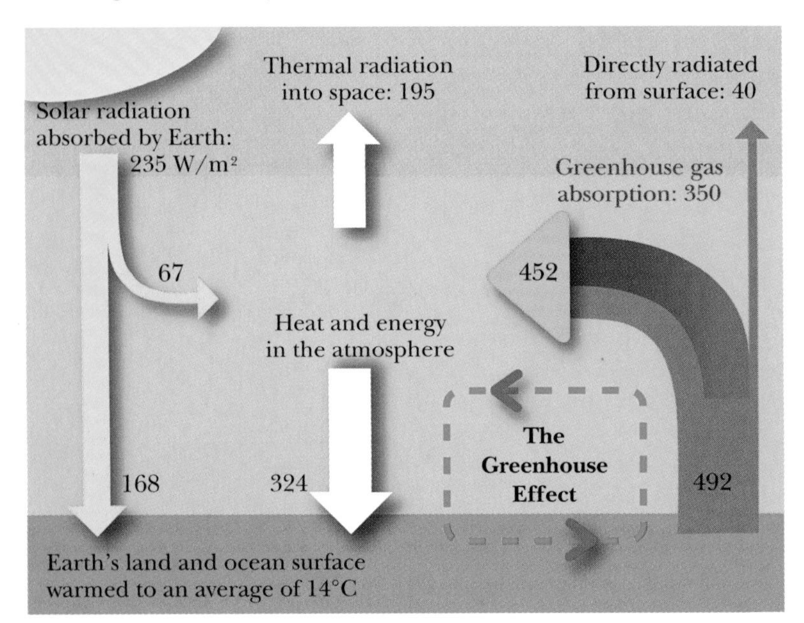

Above: The greenhouse effect is a description of how manmade or natural gases such as carbon dioxide, methane, nitrous oxide, ozone, and water vapor trap the Sun's heat in our atmosphere, preventing it from being released back into space. This diagram represents transfers of energy in watts per square meter. Top left: Power plants and oil refineries are a primary source of manmade greenhouse gases.

GREENHOUSE GASES

The phrase "greenhouse gases" immediately elicits a negative response from most people. Greenhouse gases, though generally considered bad, are in fact vital to our survival. When the Sun's heat penetrates the Earth's atmosphere, the greenhouse gases help make sure that all of it is not reflected back out into space. Some of the heat is trapped within our atmosphere, which keeps our planet at a temperature that permits living things to survive and flourish. It is a naturally occurring process. However, there must be balance, and modern industrial advances that increase the presence of chemicals like carbon dioxide and methane in the air have tipped that balance. The greenhouse gas and greenhouse effect issues are the perfect chemical example of "too much of a good thing." Although the main greenhouse gases—carbon dioxide, methane, and ozone—are produced naturally, human activities have increased their presence to alarming levels. Carbon dioxide, for example, is a by-product of plant and animal respiration. But the burning of fossil fuels has drastically increased the amount of carbon dioxide in the atmosphere. Excessive greenhouse gas levels mean more heat trapped in the atmosphere, disrupting a delicate balance that can result in detrimental effects on many of the planet's species.

TWO FACES OF OZONE

Ozone, though thought of exclusively as a layer, is actually a molecule composed of three oxygen atoms (as opposed to breathable oxygen, which is composed of two).

Ozone is formed when volatile organic compounds, such as those emitted from factories, combine with various oxides of nitrogen in the presence of heat and sunlight. This accounts for the increased likelihood of ground-level ozone pollution on hot, sunny days. Ground-level ozone exists in the troposphere—the layer of the atmosphere that extends from Earth's surface to between five and nine miles—and is a major component of smog. When particulates in the air, such as dust and dirt, combine with ozone, the results may cause increased risk of asthma, cardiovascular disease, and some cancers.

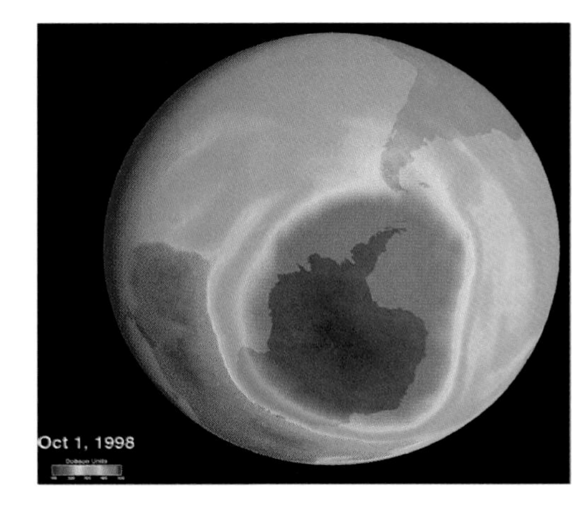

Left: This NASA image shows the hole in the ozone layer over Antarctica. Despite the hole's size, there is evidence that the ozone is slowly recovering, due to the phasing out of the use of ozone-depleting chemicals.

That's the "bad" ozone. The ozone layer, on the other hand, is not only "good" ozone, but it is essential and life preserving. The ozone layer exists in the stratosphere, which is directly above the troposphere and extends to roughly 31 miles above the Earth. It is primarily responsible for protecting us from the Sun's harmful ultraviolet radiation, which can cause skin cancer and can affect the ability of many living organisms to survive. Some of the most significant contributors to the demise of the protective ozone layer are chlorofluorocarbons (CFCs). The use of CFCs has been reduced as their effects have become better understood. However, these, along with hydrofluorocarbons, halons, carbon tetrachloride, and other ozone-damaging compounds, are commonly found in pesticides, coolants, and aerosols.

PLAYING OUR PART

Each of us does our part to add to the greenhouse gases overloading our atmosphere, heating up the planet, and damaging the ozone layer. Carbon dioxide alone is generated by many everyday activities that we take for granted. Burning just one gallon of gas, for example, releases roughly 19 pounds of carbon dioxide into the air.

Tree as pollution fighter—trees actually consume carbon dioxide and release oxygen.

TREES WITH AN APPETITE

When trees and plants take in carbon dioxide, they generate energy sources for themselves, release breathable oxygen for humans and other animals, and also reduce the amount of carbon dioxide in the air. Rising rates of carbon dioxide in the atmosphere from the combustion of fossil fuels, among other things, is a crisis affecting the health of the planet. Environmental scientists everywhere are studying the genetic makeup of trees that thrive in carbon dioxide-rich environments and using this knowledge to modify and propagate plants and trees with the capacity to take up even higher amounts of carbon dioxide. The purpose of these efforts: to reduce the global warming effects that increased atmospheric levels of the carbon dioxide are causing.

Acid Rain

Imagine a powerful destructive force raining down on the planet, slowly erasing our culture, poisoning our waterways, killing wildlife, and harming our health. It sounds horrifying, but it happens every time it rains or snows. The culprit is acid rain, a damaging by-product of our industrial age.

Acid rain, despite its name, actually begins on Earth. When we burn fossil fuels in factories, electric plants, and cars, we are spewing chemicals into the air. Two of these, sulfur dioxide and nitrogen oxide, react with water and oxygen in the atmosphere.

As it is, normal rain is already slightly acidic because it has combined with carbon dioxide in the atmosphere, and picked up CO_2's derivative, carbonic acid. Carbonic acid is weak, about 5.6 on the pH scale. When it falls to Earth in rain, it usually does not harm anything because it dissipates rapidly. (Recall that normal water is neutral, with a pH of 7.0.) However, sulfur dioxide and nitrogen oxide gain extra oxygen molecules in the atmosphere, transforming into sulfuric acid and nitric acid. When these new chemicals combine with rain and snow, they may reduce the pH of this precipitation to 4.3 or lower.

Acid rain not only harms the natural world; it can also severely damage man-made structures including buildings and sculpture. Top left: Once a forest has been decimated by acid rain, it may take hundreds of years to recover.

At this level, rain and snow can harm whatever they strike.

SUBTLE, DAMAGING EFFECTS

We may not notice the effects at first, but over time acid rain takes its toll. It weakens calcium-rich rocks such as sandstone, limestone, marble, and granite—the very same materials that make up many of our manmade structures. Bombarded by acid rain, buildings, statues, monuments, and cemetery tombstones erode and flake away. The acid seeps into the soil, leaching away nutrients and weakening trees. Chemicals build up in waterways, eradicating all life. For example, many fish eggs do not hatch in water with a pH level lower than 5; and more acidic water kills adult fish outright. The acids are even believed to damage automobile paint. Dry, particulate versions of these acids can be inhaled and damage our health.

Acid rain may seem like a modern phenomenon, but it has been around since the birth of the Industrial Revolution—scientists have observed it as far back as 1852. However, it is only during the last half century that it has become a major problem. Today, half of all the sulfur dioxide and a quarter of all the nitrogen oxide emitted in the United States come from electric power plants. The rest comes from factory emissions and automobile exhaust. That's why recent efforts to reverse the incidence of acid rain have focused on cleaning up smokestacks and exhaust pipes.

WHAT CAN WE DO?

As part of the Clean Air Act, factories and electric plants must install so-called "scrubber" technology that reduces the amount of these chemicals released into the atmosphere. In automobiles, the catalytic converter serves a similar function. One common type of converter contains a fine ceramic mesh coated with the elements rhodium and platinum that interact with emission chemicals, breaking down nitrogen oxide, for example, into its component parts, nitrogen and oxygen.

These technologies are a fine first step, but more needs to be done. The Environmental Protection Agency encourages Americans and entrepreneurs to investigate alternative fuels and energies that are less damaging to humans and the planet. Some cities are already experimenting with zero- or low-emission vehicles, for example. In other regions, biologists are teaming up with ordinary citizens to restore swaths of forestland that may have been damaged by acid rain or smog. Beyond these efforts, anything we can do to conserve energy will reduce emissions, either from automobiles or from power plants.

Emissions testing is an important element in the effort to reduce acid rain.

Radioactive Waste

One of the most sobering symbols of the modern age is the yellow, trifoil logo that warns human beings of the presence of radioactive materials. This symbol is found all over the world, in medical, industrial, military, and power plant settings. For better or worse, humans have decided to employ radioactive materials—naturally occurring or enriched chemical elements that emit dangerous particles or waves—in a range of applications. Ultimately, this means our species has the responsibility of taking out a very special kind of trash.

Radioactive waste—the disposable by-products of a variety of technical processes—is not like any other garbage the world has ever known. When we toss out household trash, those items end up in a landfill, where they are entombed forever. They may be unsightly or wasteful, but we are also reasonably certain that such materials will pose no harm to future generations. Radioactive materials are not nearly so benign. If these materials not properly handled and managed, the particles and waves emanating from uranium or plutonium can penetrate the human body, alter the sensitive chemical structure of cells and DNA, and initiate cancers and other diseases for tens of thousands of years into the future.

Nuclear power is a fact of modern life, as is radioactive waste.

Above: On August 9, 1945, an atomic bomb of the type shown here was dropped on the city of Nagasaki, Japan. Although estimates vary, many thousands of people perished after the fact from radiation poisoning. Top left: Of all hazardous waste, radioactive materials may be the most dangerous.

This places a heavy burden on any nation that creates such materials. Not only must it ensure the safety of its current citizenry, but it must also make plans to safeguard generations of citizens millennia into the future. Think of it: The planet as we know it may be very different that far into the future. The American people may speak and read a version of English that we cannot foresee. Where do we put radioactive waste so it doesn't harm those people—and how do we label it? What words do we use to warn them to stay away?

At this time in history, there are five types of radioactive waste that all nations are struggling with. In order of their danger, they are: spent fuel from nuclear power plants; transuranic waste from nuclear weapons programs and other military applications; the sandy residues from uranium mining facilities; low-level

Above: Yucca Mountain in southern Nevada is scheduled to open as a storage site for nuclear waste in 2010. It was designed to safely hold approximately 70,000 pounds of waste, or about enough to store almost all of the country's nuclear waste through 2030.

waste that includes discarded "bunny" suits, gloves, tools, and other artifacts used to handle radioactive waste; and naturally occurring waste in radioactive substance and the remains of particle accelerator experiments and commercial activities.

The half-lives of these substances—the amount of time they take to decay to less-harmful elements—varies greatly. The waste from a particle accelerator has a very short half-life; it may become inert in as little as a day. Typically, such wastes are kept on the premises of the facility until they are safe to discard. The tailings, or sandy residue of American uranium mines, on the other hand, still contain a dozen or so dangerous materials in them, such as heavy metals and radon gas, a breathable carcinogen. Uranium mining has been discontinued in the United States, because it is cheaper to buy uranium from others nations, such as Australia, which has 40 percent of the world's supply. In the United States, abandoned uranium mines have been enclosed with fencing, and are monitored. Many people are concerned that the sandy material may be blown into nearby communities by winds and storms, or wash into waterways.

Currently, the United States entombs transuranic waste—waste from materials that are heavier, by atomic weight, than uranium—at the Waste Isolation Pilot Plant (WIPP) in Carlsbad, New Mexico. This facility can hold up to 850,000 barrels of such material. The target waste, from nuclear weapons and other military applications, is primarily derived from plutonium and is considered the second most dangerous type of radioactive waste. One of the materials, plutonium 239, has a half-life of 24,000 years. Even after all that time, the waste can still harm humans.

Following decades of scientific study and political debate, the United States had planned to store the most dangerous radioactive waste—spent fuel assemblies from nuclear power plants—in an underground repository at Yucca Mountain, Nevada. At this time, the plan has been held up by numerous objections and lawsuits, chiefly from the residents of that region of Nevada. Other Americans object to the eventual transportation of these dangerous substances by train or truck cross-country to the repository site.

Recycling

Most American communities have a recycling center, where citizens cheerfully drop off household trash sorted by their materials: glass, plastics, metals, and paper. Within these categories, there are subcategories, some easier to comprehend than others. Everyone can tell brown glass from white glass. But metal from aluminum? A No. 1 plastic from a No. 2 plastic? Mixed paper from cardboard? These are the fine distinctions that trip up well-meaning people every time. We may not think these distinctions matter much, but for some objects it matters all the way down to the molecular level.

Over the last two decades, Americans have grown familiar with the recycling logo (top left) as well as sorting paper, aluminum, glass, and plastic for recycling. Below: Once it has been mulched, nontreated wood can be reused as an effective top layer that holds heat and moisture in the soil of a garden.

Recycling takes advantage of what scientists call the law of conservation of mass. First expressed by the French chemist Antoine Lavoisier (1743–94), this law says that though matter may change its shape or form, it cannot be destroyed or created. When we recycle, we are transforming discarded matter by changing its shape, and possibly its properties and applications. We are taking an aluminum soda can, melting it down, and turning it into siding for a home, a piece of electronics equipment, a cookware pot, or yet another soda can. The polymers in a plastic soda bottle may be recycled into a comfy fleece blanket or jacket. Recycled matter is not and cannot be destroyed. It simply changes its form.

Recyclables are just a start. Americans never thought twice about sending "green waste" such as grass clippings, brush, and tree trimmings to a landfill. Now cities urge homeowners to chop up this material and use it to mulch their flower beds. Mulch, a thick layer of organic woody or grass material, protects plants by absorbing rainfall and shielding delicate roots from the harsh Sun. Today, a new product called mulching mowers chop up grass blades finely and scatter grass particles on lawns, which are fertilized as the dead grass breaks down into its essential nutrients.

Some materials, however, are hard to recycle. Plastics are complex polymers that mix smoothly only with polymers like themselves. Dissimilar plastics don't mix; they separate like oil and water. That is why it is so important to correctly sort different types of plastics at a recycling center. It is also why some centers accept only certain types of plastics. The recycling companies that partner with communities may only be able to process and remanufacture plastics with a resin identification code of 1 or 2. It does no good to toss in plastic items labeled 3 or higher, because they cannot be used, will not mix well with the first two plastics, and will only have to be discarded. Most communities recycle only plastics 1 (PETE or PET), 2 (HDPE), and 3 (PVC) because these currently are the most cost-effective polymers to recycle. In the United States, a plastic's resin identification code is stamped on the item in a rounded triangular logo that encloses the plastic's number.

Scientists are constantly devising new plastics recycling methods, and it's possible that someday all plastics will be readily recycled. One technology, called froth flotation, makes it possible to sort two of the most common and valuable plastics: acrylonitrile-butadiene-styrene (ABS) and high impact polystyrene (HIPS) from a mixed junk stream of other mixed plastics. The method could recover 300 million pounds of plastic each year in the United States alone, and eliminate 10 trillion BTUs of the nation's energy consumption because new plastics would not need to be made. Following Lavoisier's law, we would simply change the form of the old plastics.

Electronic waste, or e-waste, from computers, telephones, and other electronic equipment is piling up. That material can contain chemicals and metals that are hazardous to the environment. To combat the problem, many manufacturers and local governments now offer advice about recycling these items.

ELECTRONIC TRASH

Computers, cell phones, digital cameras, mp3 players—the list of electronics available to consumers seems to grow exponentially. So, too, does the trash generated by the disposal of these modern-day "necessities," as consumers change their gadgets sometimes more often than the tires on their cars. Where does it all go? Five to seven millions tons of these devices become obsolete every year, ending up in the trash. If current recycling rates stay the same, only about 10 percent of this high-tech trash will be recycled. Known carcinogens like the elements lead, mercury, and cadmium found in batteries, circuit boards, and computer screens could potentially leach out of landfills and into soil and groundwater. Some states have already banned electronics with screens from their landfills.

CHAPTER 11

LOOKING FOR CHEMICALS

Scientists study the compounds found in water, before its use, in a lab (left), and after its use, in a wastewater sedimentation tank (top), to determine how the water is used and what changes it. Bottom: The earliest recognizable chemists, alchemists, likewise studied changes, hoping to discern what could possibly be happening at scales they could not perceive.

For centuries, scientists have dreamed up myriad ways of exploring the mysteries of chemicals and their constituent elements and molecules. These chemical entities are often extremely small and cannot be seen without the help of technology. Imagine: An elusive, chemical presence may exist in an unknown gas or liquid. The structure of a tiny organism cannot be seen with a regular light microscope. To surmount such obstacles, scientists have developed an array of methods for studying these elusive subjects, and though technology helps, creativity is key.

Scientists have devised means of identifying elements by studying their behavior and interaction with other compounds, by examining what they can and cannot do, or by analyzing a particular energy that they emit. Analytical chemistry seeks out the elements that have gone into hiding. It does this in many different ways: by exploiting the manner in which atoms behave among each other, by determining how they are affected by magnetism and light, and by tracking the behavior of their electrons. The only thing standing between scientists and the untold secrets of chemical compounds is imagination.

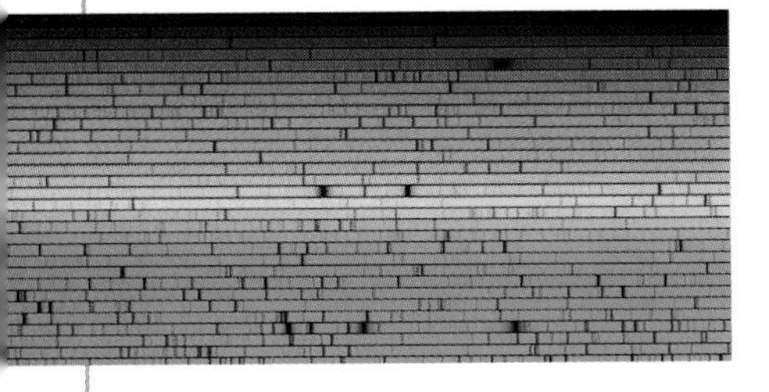

Spectroscopy

The light that shines from the Sun arrives on Earth and illuminates the structures we see around us. But every ray of light has a highly complex composition that we can't see without the help of scientific instruments. Differences within the spectrum of visible light are familiar to anyone who has ever held a prism up to the Sun or witnessed a rainbow after a storm. Other forms of light energy invisible to the human eye can be used as a means of identifying and differentiating between different substances. The study of these spectra, these different light energies, is referred to as spectroscopy and has been going on since the 1800s, though the tools and techniques have evolved over time.

ABOUT LIGHT

Light travels as a wave, and is often thought of as similar to waves on an ocean. Waves have a certain height, or amplitude.

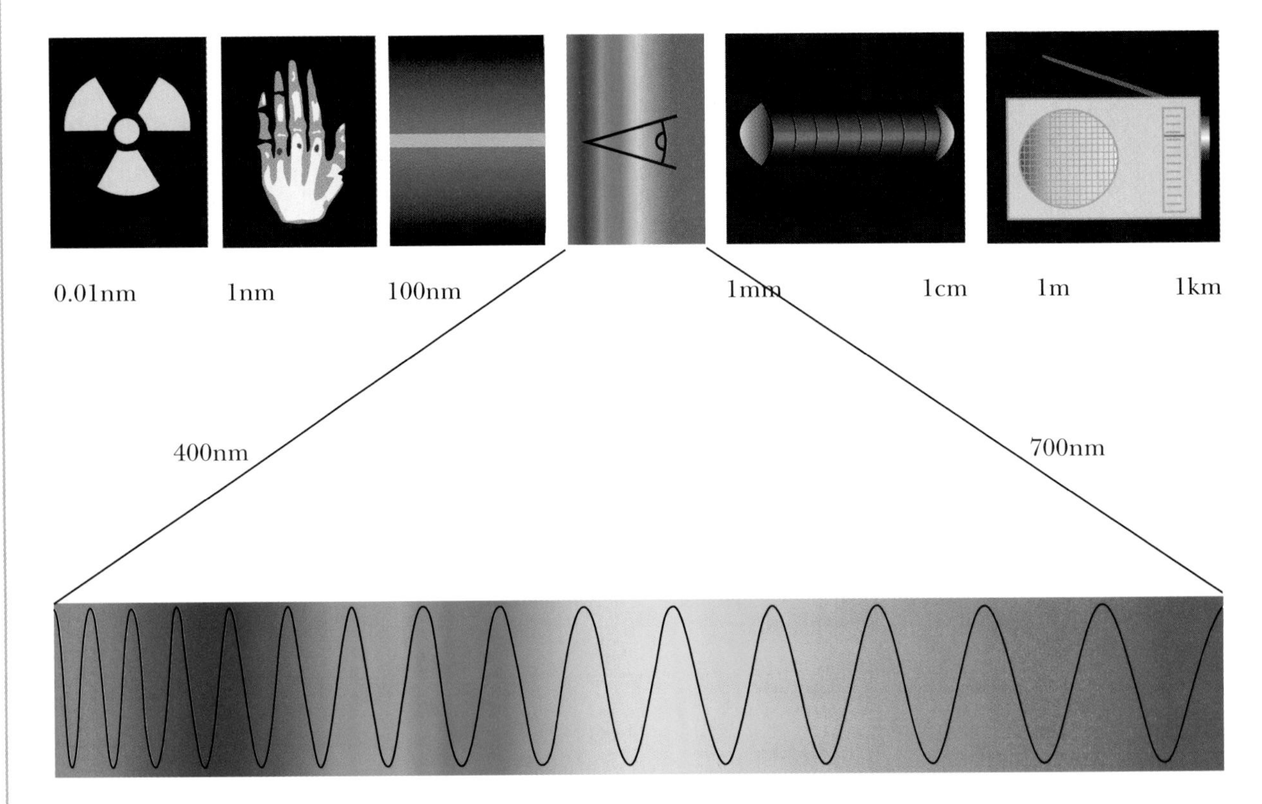

Above: The light we see constitutes a very small part of the electromagnetic spectrum. At wavelengths longer than that of red light are the infrared and radio spectra. Wavelengths shorter than violet lead into the ultraviolet spectrum, and then into dangerous radiations such as X-rays and gamma rays. Top left: The dark bands represent the wavelengths of light absorbed by atoms between the source and the spectrometer.

Atomic absorption spectroscopy can detect heavy metal particles in concentrations as low a one part per million.

These waves strike objects with a specific frequency and behave a certian way. The waves also travel at specific speeds—the distance between crests of successive waves is referred to as the wavelength. The electromagnetic spectrum classifies different kinds of light according to their wavelengths or energy levels, since light is, after all, a form of energy. Simply put, the longer the wavelength, the lower the energy, and vice versa.

BREAKING IT DOWN

Materials both emit and absorb light. In fact, every single chemical element in the universe emits and absorbs light at very specific wavelengths in a pattern that is unique to that element. Oxygen atoms, for example, emit green light when excited. The emission and absorption of light takes place at the atomic level. As the electrons within an atom move between energy levels, they lose or gain energy in the process. An electron that drops to a lower energy level—that is, one closer to the nucleus—will lose energy. The energy that is lost is emitted as a photon, or a distinct package of electromagnetic energy. The wavelength of the photon depends on how big or small a jump the electron has made: If an electron drops just one level, it will emit light of a longer wavelength and lower energy. If the electron were to drop two levels, it would emit light of a shorter wavelength and higher energy.

Since the relationship between these electron levels is specific to the kind of atom, the photons emitted or absorbed in the process tell a lot about the atom or molecule being examined. Spectroscopy uses information gained from the emission and absorption patterns of materials to identify and better understand their makeup. It is therefore especially useful in analytical chemistry and has long been used by astronomers to determine the composition and characteristics of different celestial objects.

DIFFERENT WAVELENGTHS FOR DIFFERENT JOBS

There are various kinds of spectroscopy, including Raman, infrared (IR), electron spin resonance (ESR), nuclear magnetic resonance (NMR), and gamma-ray and microwave spectroscopy, among many others. Each has its own strength and is used for specific applications. For example, infrared spectroscopy, which uses infrared light, is particularly useful because virtually every molecule has an infrared spectrum. Infrared spectroscopy also helps in identifying the type and strength of chemical bonds present in materials, since the molecules will vibrate at distinct frequencies depending on what kind of bonds they contain and how many, as well as their orientation.

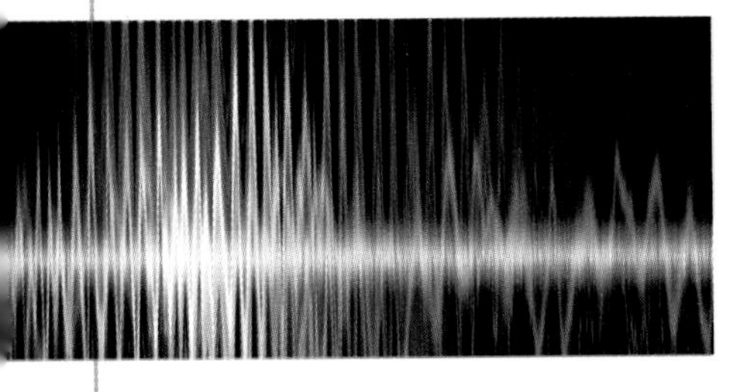

Nuclear Magnetic Resonance and Magnetic Resonance Imaging

Today doctors are able to inspect living tissues without destroying them, and study the internal workings of living patients without the benefit of exploratory surgery. We may be familiar with the terms "MRI" or "NMR," but chances are we do not realize that doctors are exploiting the tiny magnetic fields carried by certain atoms to glean glorious new pictures of the human form. Here's a look at the technology they use.

NMR

Some atoms contained in cells carry an odd number of protons and neutrons, which gives their nuclei a slight magnetic field. When scientists place a sample of these cells in the center of a superconducting magnet, a few of these little atomic magnets line up along the north-south axis of the external field. Next, if scientists blast these atoms with radio waves at the right frequency (the "resonant frequency"), the atoms will absorb energy to become excited, deviating momentarily from the prevailing magnetic direction.

As they fall under the big magnet's sway again, the atoms radiate the waves back out while an antenna registers and charts the resonance. Every atom has a highly specific, signature frequency, although this frequency is affected by neighboring atoms. Scientists can then use these frequencies to identify exactly what elements are present in the cell and infer how the elements are grouped into molecules, the structure of those molecules, and how they are folded up to fit inside the cell. NMR only works on certain isotopes, or variants, of hydrogen, carbon, nitrogen,

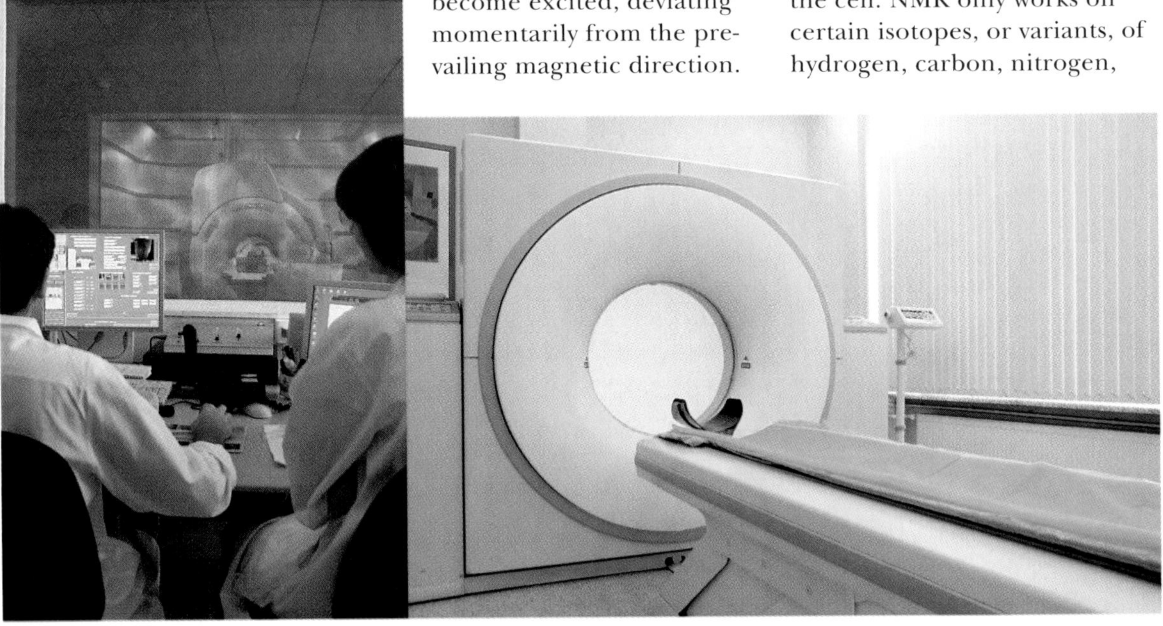

Above: Technicians use computers (left) to control the massive magnets used by NMR/MRI machines to take accurate images of the inside of a person's body. Top left: Radio waves are a pivotal component of resonance, which drives nuclear and magnetic imaging technologies.

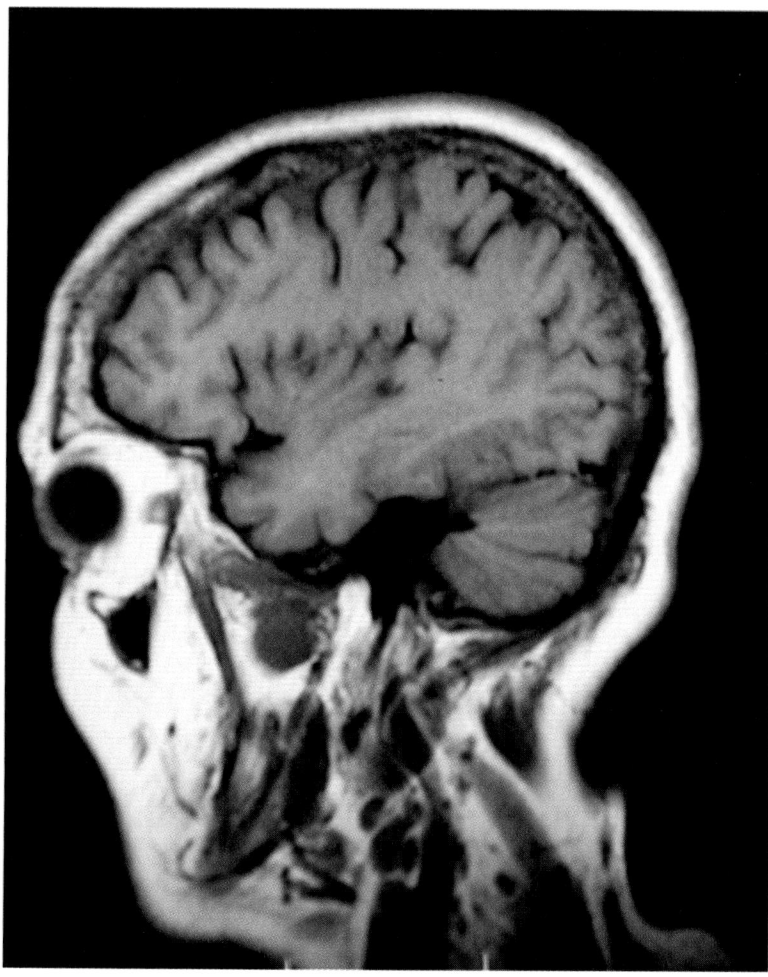

A single slice of a full three-dimensional NMR/MRI scan of a human head.

and other elements, which just happen to be abundant in living cells. Today's NMR devices tend to focus on hydrogen, the most abundant element.

The magnets used in nuclear magnetic resonance exert forces that are stronger than the pull of the Earth's magnetic field and technicians who work with this equipment must take special precautions. Such magnets have been known to stop watches, erase credit cards, and even disrupt pacemaker devices implanted in peoples' chests. The stronger the magnet, the greater the resolution, or detail, will be in the final image. Edward Mills Purcell (1912–97) and Felix Bloch (1905–83) received the 1952 Nobel Prize in physics for independently establishing the scientific principle behind NMR.

MRI

Building on the discoveries of Purcell and Bloch, later scientists were able to develop a giant magnet capable of scanning the entire human body. The first magnetic resonance imaging (MRI) scanner was tested on a human subject in the 1970s. The test took five grueling hours. Today the same test takes only seconds, although average scans may run as long as 20 minutes or so. The patient lies still on a flat table that slowly moves into the center of a giant magnet. From there, the magnet behaves very much like the NMR device described earlier. The magnet focuses on one thin "slice" of the patient's body at a time. Hydrogen atoms in the patient's body yield clues to the patient's physiology, which are assembled by a computer into coherent, two- or three-dimensional pictures that can be read and interpreted by physicians.

Although MRI scans are quite costly, they make it possible for doctors to see the interior of the body—the brain, liver, kidneys, everything—in 3-D without radiation. The MRI images very clearly differentiate between muscles, vessels, fats, cartilage, and bone. They can show, for instance, if a patient has a blockage in an artery of the heart, or the brain. In 2003, scientists Paul Lauterbur (b. 1929) and Peter Mansfield (b. 1933) were awarded the Nobel Prize in medicine for their work in developing MRI technology. MRI machines are constantly evolving and improving; the most modern equipment tends to be lighter and less claustrophobic.

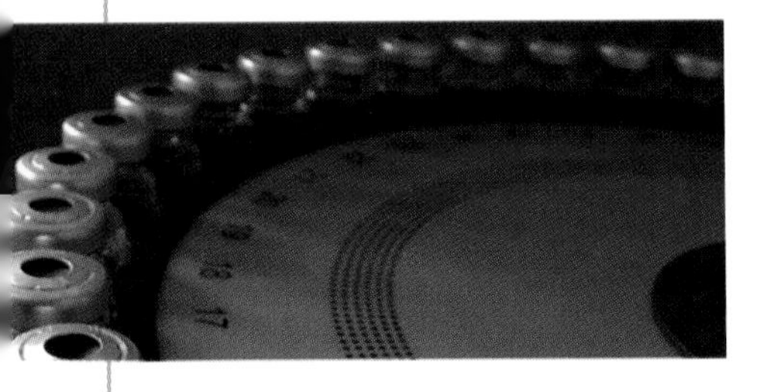

Chromatography

Imagine an errant drop of water landing on a note that was handwritten in black ink. Fifteen or so minutes later, the ink appears not only to have spread, but there is a ring extending out from where the original drop fell. Surprisingly, the color of that ring is not just black, but also blue and even a bit of what appears to be red. This phenomenon is chromatography at work: The different components of a mixture—in this case, ink—interacting with a medium (paper) in the presence of a solvent (water) and separ-ating according to their chemical properties.

From the first time Russian scientist Mikhail Tsvet decided to separate out plant pigments in 1901, chromatography has become a valued tool for chemists seeking to analyze unknown mixtures and gain insight into their composition. Mixtures pervade our world. The foods we eat are largely mixtures, composed of a variety of different compounds, each possessing its own chemical personality. Mayonnaise, as we've seen, is a mixture classified as an emulsion, whereas a nice oil-and-vinegar dressing is a mixture described as a suspension. Many fields of analytical chemistry and biochemistry are dependant on the ability to analyze and purify the different components in a mixture. Separating these mixtures into their individual components can be done in many different ways, including simple filtrations as well as more advanced techniques like chromatography.

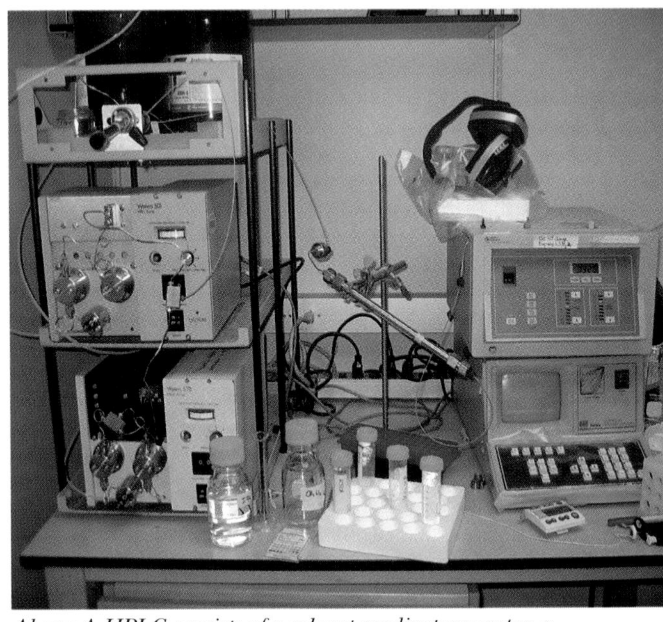

Above: A HPLC consists of a solvent-gradient generator, a steel-reinforced reaction chamber, and a computer for analysis. Top left: Vials arranged for processing in a gas chromatograph.

Dyes and inks are often mixtures of many colors that can be separated using chromatography paper.

SEPARATION AND IDENTIFICATION

Chromatography is a means of analyzing the components of a mixture and also their concentrations. The mixture to be analyzed is aptly referred to as the analyte. The analyte is combined with a solvent, which may be liquid or gas. Together, these are called the mobile phase. The two will travel together through a stationary phase—a piece of chromatography paper, for example. Individual compounds present within the mixture being analyzed will interact differently with the solvent and with the stationary phase. So, as the compounds move along, some will be more attracted to the stationary phase, others less so. They will therefore travel along the stationary phase at different rates. Imagine the mixture to be analyzed is a group composed of four different students, and the stationary phase is rope they have to climb in gym class. Each student will reach the top of the rope in a different amount of time, and no two will pause along the way at exactly the same spot. How far the different components of a mixture travel and the amount of time it takes them to do so are directly related to their atomic and molecular structure.

There are many different chromatography techniques that are used today, and the technique, solvent, and stationary phase used depend on the nature of the mixture being tested. Two very common forms are gas chromatography and liquid chromatography. In gas chromatography, gas is considered the solvent for the mixture being analyzed.

In liquid chromatography, a common stationary phase used is silica. The silica is treated with different compounds depending on the mixture that is going to be analyzed and then is packed inside a tube. The mobile liquid mixture gradually makes its way through the stationary silica, and individual compounds get "stuck" along the way, leaving a kind of chemical fingerprint that scientists can use to identify them. A highly dependable form of liquid chromatography in use today is called high-performance liquid chromatography. HPLC uses higher pressure to achieve greater efficiency.

APPLICATIONS

Chromatography is highly effective and precise, capable of separating out different compounds that possess even the most subtle of differences. Chromatography can be used for pure, empirical research, as a means of better identifying the chemical makeup of an unknown mixture and clarifying the concentrations of its various component compounds. And chromatography also has many applications in the fields of biotechnology and analytical chemistry. Both gas chromatography and high-performance liquid chromatography are used to help determine the levels of mercury in seafood. Chromatography can also be used to determine the kind, and concentration, of specific contaminants in a given sample of drinking water. And chromatography is indispensable in drug testing, and is used to test for the levels of drugs such as cocaine in urine and blood.

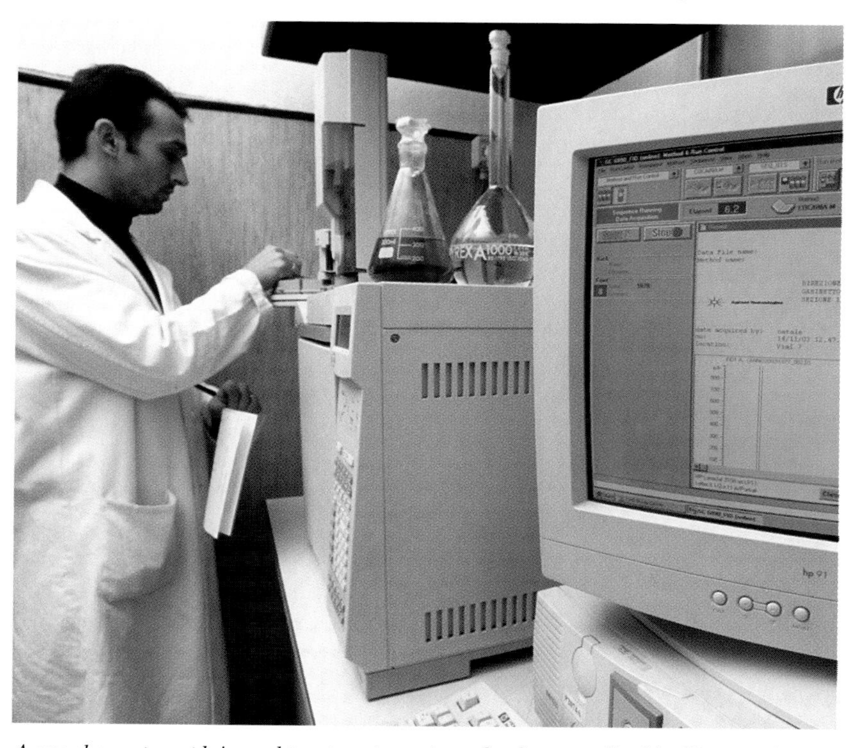

A gas chromatograph is used to separate compounds of gases or liquids. Here, a scientist prepares to analyze a sample. The computer screen will display the results of the analysis.

Electron Microscopes

Who has not peeked through the lens of a microscope at least once in his or her life? The most popular type of microscope uses glass lenses to observe small objects placed on a glass slide. By adjusting the controls, the observer can focus the lenses to magnify the object. Because they work by bending light, such microscopes are called light microscopes . Although these microscopes are excellent, they are also quite limited in their abilities. They can magnify objects by only 500 to 1,000 times. Because this is not powerful enough to help many scientists with their work, they rely on a different kind of microscope altogether.

Above: The precision stage on which samples are placed in a scanning electron microscope. Top left: Control station of a scanning electron microscope.

A scanning electron microscope (SEM) does not use light and glass lenses to magnify objects. It instead works by tracking the movement of electrons. An electron gun at the top of the microscope shoots electrons down a long tube, or column. The electrons travel down along the column, where they are then focused by electromagnets into an electron beam. The beam shoots out of the column and bombards the specimen that is being studied. The electrons penetrate the orbits of the millions of atoms that make up the specimen, and behave in one of two ways. Some electrons bounce around the nuclei and depart the atom in a different direction. Some electrons whiz past other electrons, causing them to shoot out of their orbits. Specialized detectors spot these electrons as they fly around, and "read" their presence as an electric signal. This signal is sent to the television screen that is connected to the microscope, which converts the signal into an image. Incidentally, this technology is not unlike the inner workings of an ordinary television, which converts an electron beam into pictures of your favorite celebrities.

The term "scanning" refers to the microscope's ability to intelligently collect information in one spot and transcribe that information in another spot. First, the electron beam scans the specimen horizontally and vertically. Then it transmits that information to the computer, which transcribes the same data on the television screen for the scientist to view and interpret. If the scientist wants to increase the magnification, he or she must scan a smaller portion of the specimen. The smaller the area being scanned, the larger it will appear on the screen. Consider the zoom feature on a standard camera: If the photographer zooms out, he or she sees more of the scene, but with less detail. On the other hand, zooming in will produce an image in which a person's face will be magnified and that person's body will not be visible. Electron microscopes work in much the same way, but they can magnify samples to more than 200,000 times their actual size. That's a huge leap from light microscopes!

Electron microscopes were first developed in the 1930s and 1940s but were not commercially available until 1965. Today they are standard equipment in most university laboratories. And yet, for all their advantages, some models still have a central shortcoming: specimens need to be coated with gold foil before they can be scanned. This makes them strong electrical conductors, and thus easier to see. Many modern scientists also rely on nuclear magnetic resonance (NMR) technologies, which measure changes in the magnetic fields of atoms that have been exposed to radio waves, to explore the inner workings of living tissues.

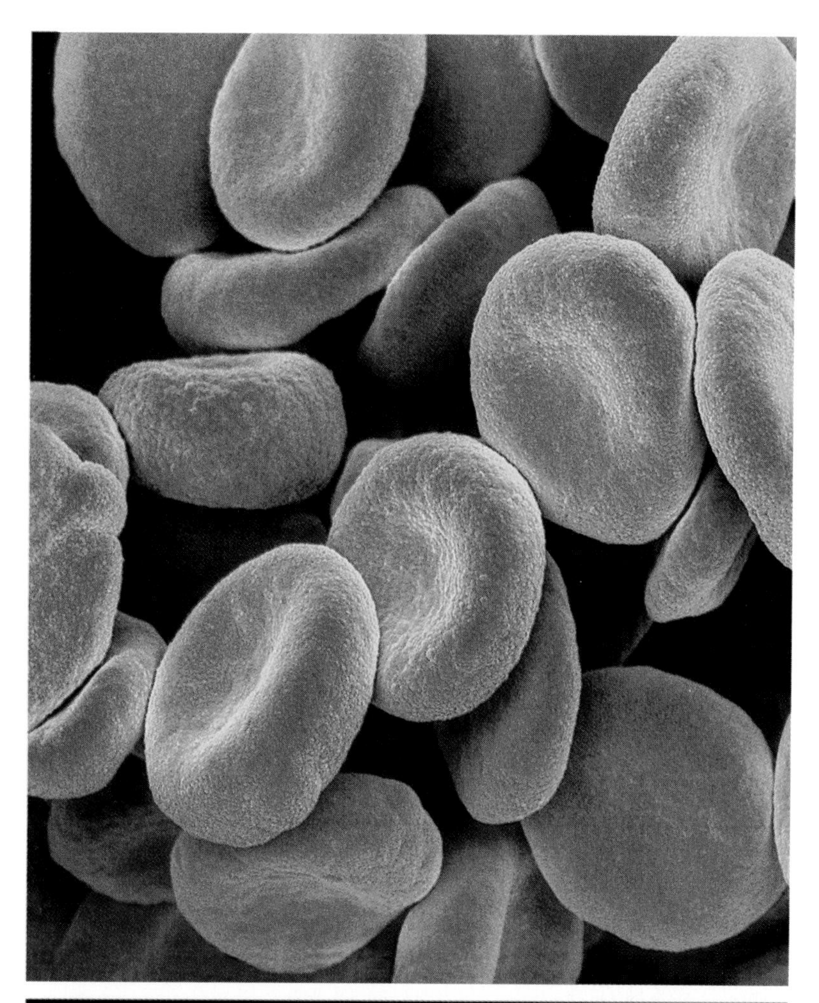

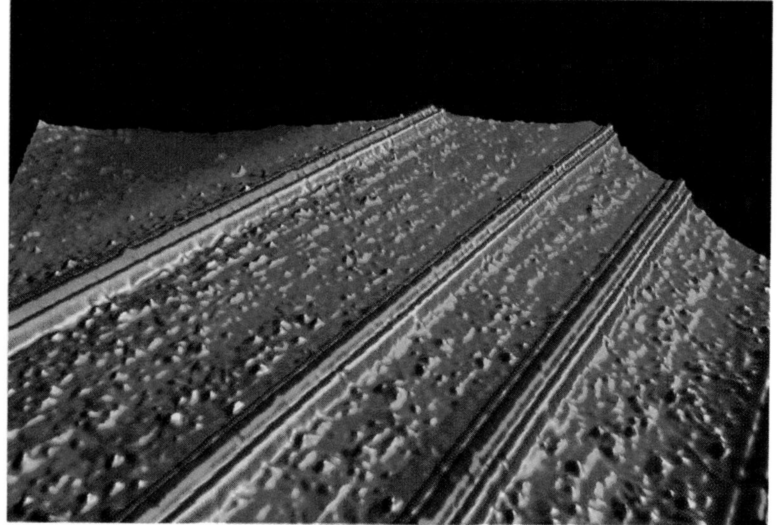

Top right: Human red blood cells as seen through an electron scanning microscope. Right: Red nanowires contrasted against a blue substrate in a color-enhanced surface scan.

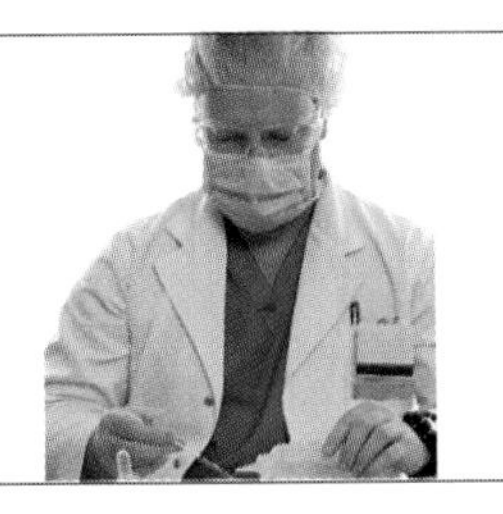

Forensic Chemistry

In recent years, few scientific fields have captured the imagination of a general audience more than forensic science. Television has been inundated with prime-time characters in designer clothes, who populate sparkling labs doused in electric-blue light. Though the flashy togs, centrifuge montages, and soundtracks may be absent from the average federal crime lab, the chemistry that these fictional characters has popularized is quite real. Forensic chemistry is the star of many real-life crime investigations—although it usually takes a lot longer to analyze results in the real world.

SCIENCE AND THE LAW

The word "forensic" in and of itself does not have anything to do with science. Something described as forensic means that it relates to legal proceedings, public debate, or courts of law. There are forensic anthropologists and forensic psychologists. So, forensic science is science that plays a role in the legal system, and this does often include criminal cases. Outside of the lab, forensic chemists specializing in DNA analysis may have to testify as to the results of their work. The field of forensic chemistry is usually highly specialized. There are forensic chemists who focus their talents and abilities on nothing else but identifying illegal substances, for example. Forensic chemists work in a variety of capacities and depend on a number of different analytical methods to help them find the answers they seek. Some methods are more familiar than others, and as new laboratory procedures are developed and advances are made in analytical chemistry, an increased arsenal of forensic tools will be made available.

Above: An infrared spectrometer can be used to look for certain chemicals, like drugs or explosives, and can also determine what else might be mixed with those substances. Top left: Forensic chemistry is a highly specialized field. A forensic chemist may specialize in specific laboratory tests.

DNA ANALYSIS

Everyone's DNA is unique, and it can be identified from even the most minuscule of bodily samples—a hair from a head, saliva from a sealed envelope. DNA is routinely used to identify the guilty and, now in increasing instances, to free those who were wrongly incarcerated in the past. Enzymes can be used to make hundreds of thousand of copies of a small sample of DNA. Enzymes can also be used to cut the DNA so that specific patterns that identify only one individual can be highlighted.

CHEMISTRY AT THE AIRPORT

Many airport security checkpoints can now benefit from advances in forensic chemistry. In some airports, security officers will wipe down laptop computers or other pieces of luggage and then insert the swab into a machine while the passenger waits for just a few moments to see whether or not they can proceed. Once the swab is inserted in the machine, the organic compounds it picked up are evaporated and mixed with a gas, and the whole mix ends up in another compartment, where it encounters high-energy electrons emitted from radioactive nickel. The result of all this? Ions. And what these machines do next is examine the behavior of these ions as they travel through gas in an electrical field within the machine. This helps determine their relative size. Basically, larger ions travel slower than smaller ions because they are heavier and collide with more gas molecules along the way. The machine analyzes the results and determines if they match the profile of any dangerous substances. This technique is called ion mobility spectrometry.

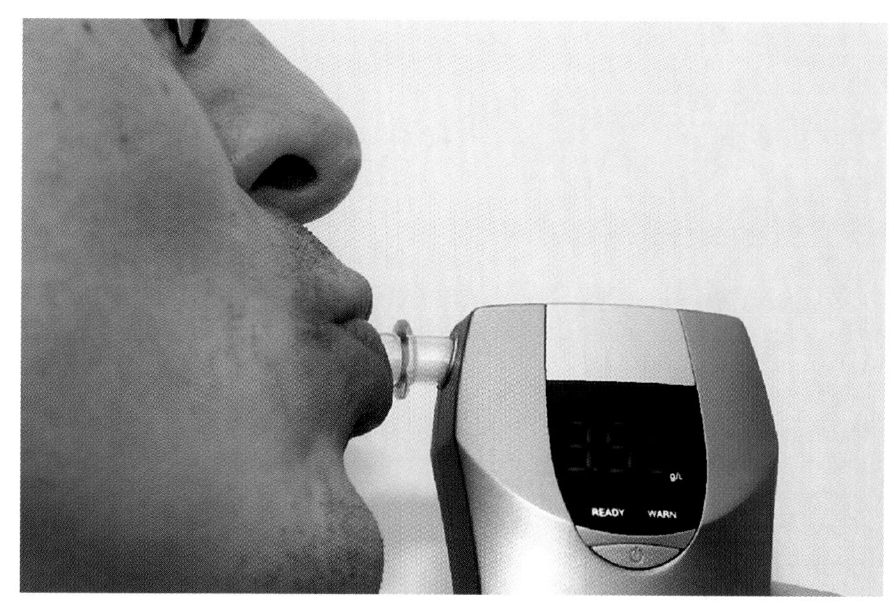

A Breathalyzer test detects trace amounts of alcohol being released from the bloodstream into the lungs.

CRIME SCENE

Forensic chemistry is also useful in cases of arson. High-performance liquid chromatography and mass spectrometry can be used to analyze the debris from a fire to determine whether any chemical compounds present are specific to known fire accelerants. If such a substance has been found, the pattern of hydrocarbons in the sample could be matched to a potential source.

Forensic chemists help federal officials follow the trail of counterfeit money, by using gas chromatography and mass spectrometry to identify and analyze different kinds of inks.

Forensic toxicology is commonly used to examine the levels of drugs and alcohol in body tissue, living or dead, and relies heavily on gas chromatography to separate out and analyze the levels of concentration of various chemical compounds. The peaks and valleys that show up on a chromatogram can tell chemists the concentration of alcohol in a blood sample, and whether chemical compounds present in the blood, urine, or tissue match the chemical signature of known illegal substances.

PET DETECTIVES

A rare species of bird is found shot illegally in a protected area. Who investigates? Nature's crime scene investigators are scientists in an expanding area of forensic science called wildlife forensics. One of the major challenges for these scientists lies in the nature of their victims. Instead of dealing with one primary species—humans—they are examining evidence taken from a wide range of critters. DNA analysis plays a major role in the wildlife forensics crime labs as well, and can help with species identification among other things. One of the aims of wildlife forensics is the investigation of violations of wildlife protection laws, including poaching.

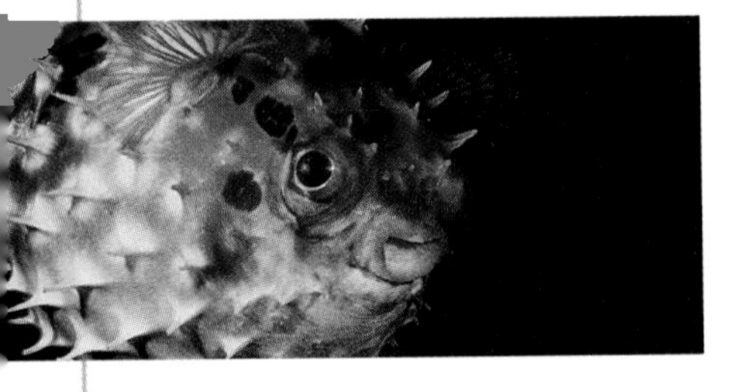

Mapping the Genome

In the year 2000, an international team of scientists—based in the United States and the United Kingdom—announced that they were close to deciphering the 3.1 billion "letters" in the human DNA molecule. Remember: The sides of a DNA molecule are held together by a seemingly endless sequence of nitrogen base pairs—A always pairs with T, and C with G. To sequence DNA, scientists spell out the precise order of every DNA base in the genome. To speed up their work, they sequence only one side of the molecule. At the same time, they make a genome map of their findings. A genome map is less detailed than a sequence. It does not show the individual bases but instead shows where on a strand of DNA a particular gene is located. A gene could be several thousand bases long. It is possible to map a gene, to know where it is located in DNA and what it does, and still not know its sequence.

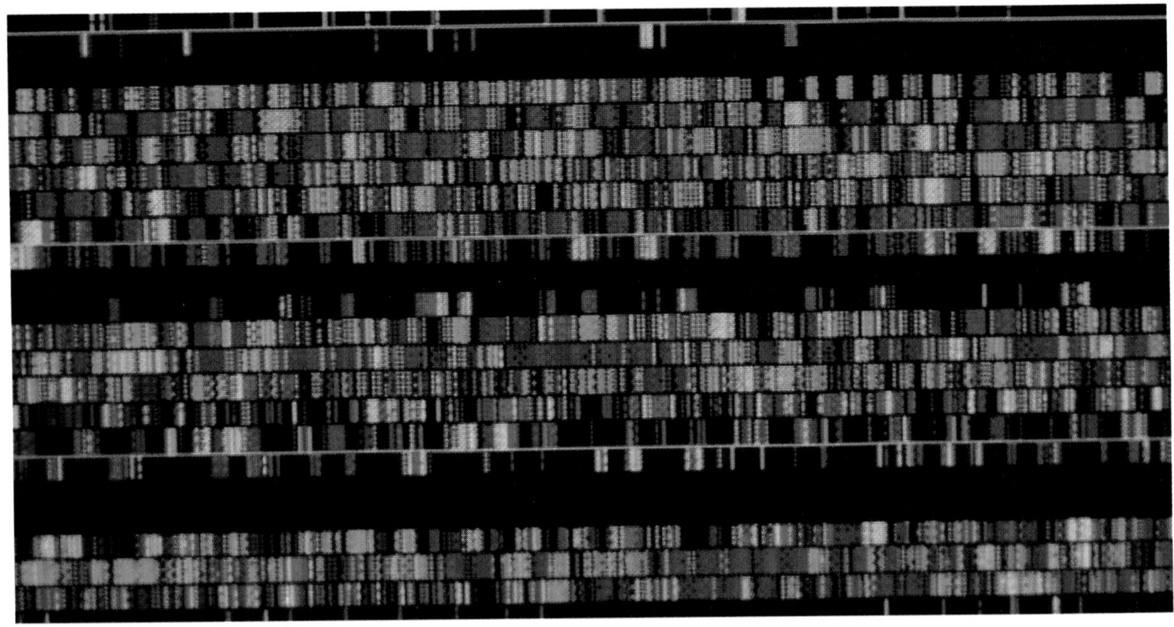

A color-coded map of the human genome, with each base pair represented by a different color (above) and physician-geneticist Francis Collins (above right). Top left: A puffer fish has a peculiar self-defense mechanism, dictated by its genetic lineage.

To understand the complexity of their task, consider that human DNA contains about 30,000 to 40,000 separate genes. Only 1/60 of our DNA contains information to make proteins. The rest tells RNA when and where in the body each gene should be used. Although 99.9 percent of any two humans' DNA is identical, that final 0.1 percent variation allows for differences in the way people are affected by disease or environmental pollutants and in how they respond to drug treatments.

The race to sequence the genome took about 15 years and enlisted the aid of scientists working as part of the Human Genome Project, a U.S./U.K.-based effort, and private corporations, most notably Celera Genomics, a Rockville, Maryland, company. Competition between private and public forces helped speed up the DNA sequencing. The private team, Celera, led by biologist J. Craig Venter (b. 1946) used a time-saving method, known as shotgun sequencing, to break up lengths of DNA into smaller pieces, analyze them, then stitch the results together. Celera's method, first decried as sloppy and error-prone by other geneticists, was eventually adopted by the Human Genome Project under physician-geneticist Francis Collins (b. 1950) because it proved quicker and cheaper than methodically sequencing one base at a time down the length of DNA. The cost of sequencing the genome also dropped markedly in the course of the project, from $10 a base at the start to nine cents per base at its conclusion.

That private corporations or individuals might patent all or part of the genome has been a concern from the earliest stages of the project. Most scientists agree that the genome should remain free to all, since it is part of our shared human heritage on Earth. It is not clear, even at this date, what will happen to the patents that many companies have applied for as they have identified various genes.

The patents have not been granted, as there are still vast ethical genomic issues to be considered. Is it ethical, for instance, for a health insurance company of the future to deny patients coverage based on the content of their DNA? Is it ethical for parents to knock out, or eliminate, genes from their DNA to spare their children inherited family conditions or diseases? If so, can those same parents choose DNA that will guarantee their children blue eyes, superior intelligence, and above-average height? Such questions remain to be answered as humankind faces the implications of unlocking the secrets of the genome.

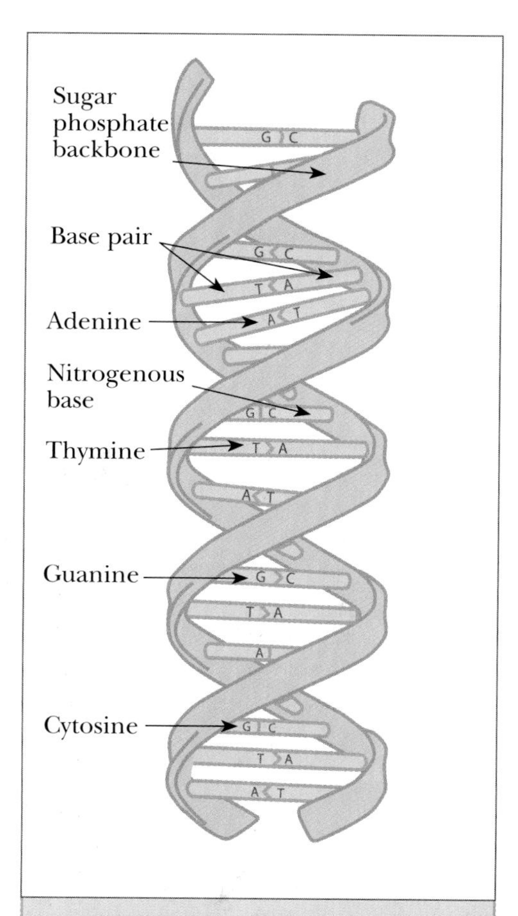

A simple rendering of the double helix DNA molecule shows the sugar backbone and base pairs.

HOW MUCH IS THREE BILLION?

There are 3.1 billion base pairs in the human DNA molecule. That number sounds large, but most of us have no idea just how large it is. Understanding DNA's complexity—the endless stream of molecules alternating between A-T and C-G in a long, unbroken chain—can help us understand why it took scientists 15 years to get the letters straight. Imagine this: If a person typed all the letters present in the entire genome sequence—A, T, C, G, and so on—at the same size that they appear in this book, he or she would produce a stack of 8 1/2 X 11 paper nearly as tall as the Washington Monument!

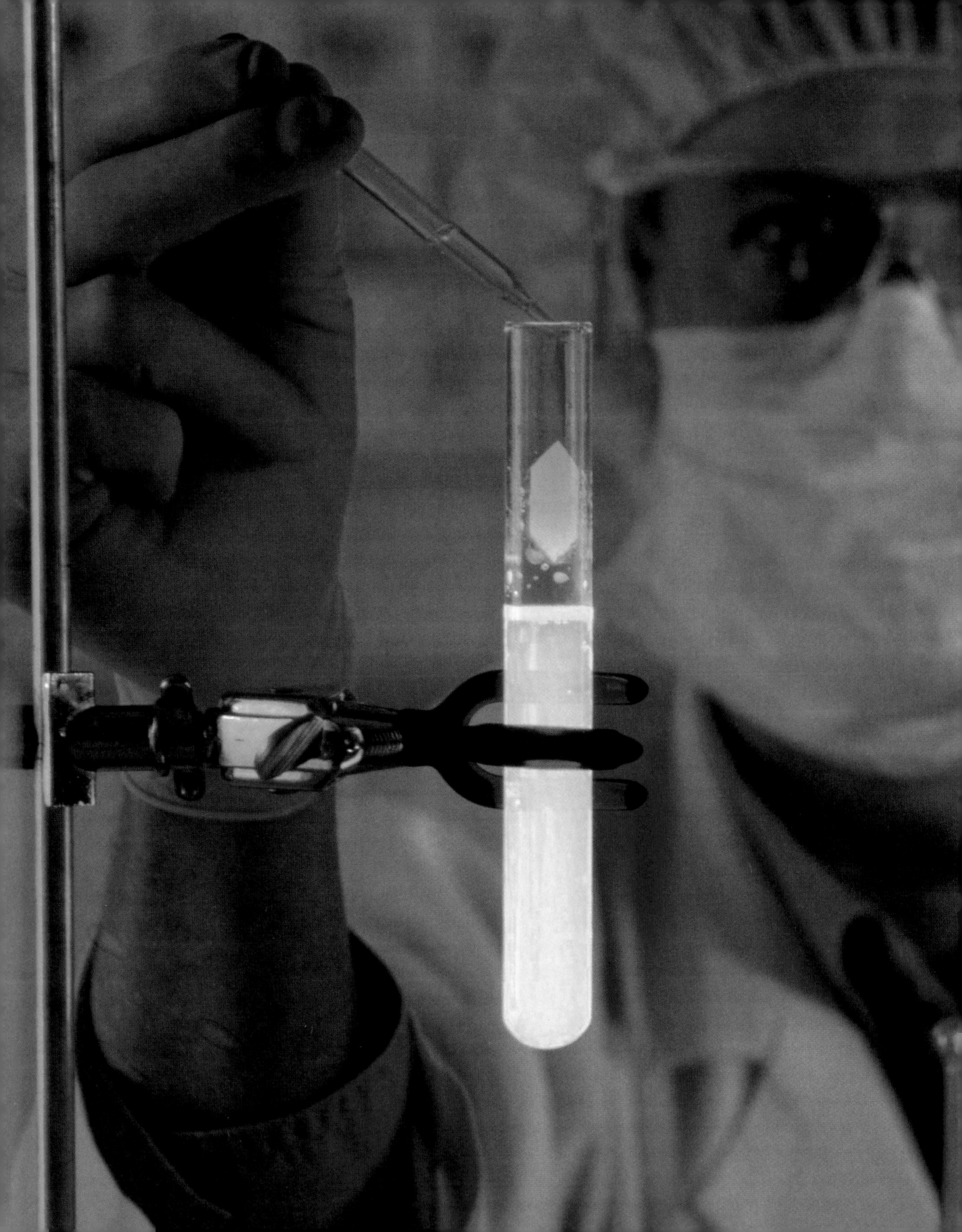

CHAPTER 12

WHERE DO WE GO FROM HERE?

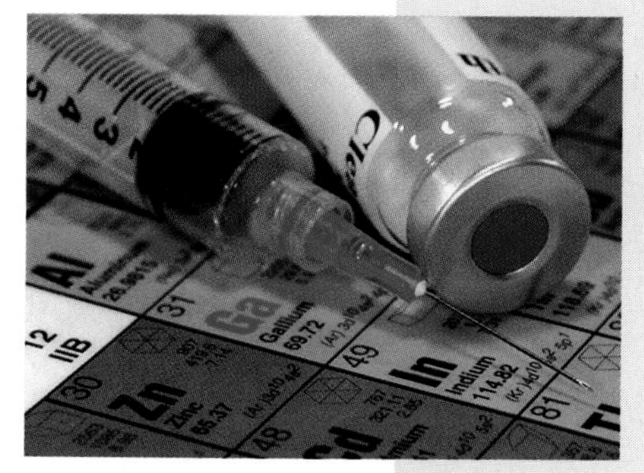

Left: Advances in chemistry have led to groundbreaking developments in numerous fields, including medicine and material sciences. Top: Nanotechnology is an interdisciplinary field that deals with the control of materials on the atomic level. Scientists believe that the miniaturization of computer processing units (CPUs) will greatly benefit the fields of biotechnology and computation. Bottom: Researchers are continually testing for drugs to fight new, resistant bacteria strains.

The challenge and the thrill of chemistry, and its virtually limitless applications, lie in the fact that the game is always changing: There is no end, no final achievement, no culminating discovery. Everything new is old again, in a sense, as scientific advancements quickly render obsolete the remarkable accomplishments of a previous era. Each new discovery feeds the progress of a different discipline. Advances in computer technology, for example, have refined researchers' abilities to "see" molecular structures, allowing for improvements in drug-designing methods. Nanotechnology is taking computer-processing units down to the scale of the atom, and someday we may have functional computers that can travel through the vascular system, monitoring our health and delivering drugs. Yet nature trumps all. Infinitely adaptable and always unpredictable, nature often lobs a curveball at us—in the form of a resistant strain of bacteria, for example—just when we think we have created the perfect antibiotic. Yet this same source of frustration is so often the grand inspiration for some of mankind's greatest scientific endeavors. Nature throws down the gauntlets and science gladly accepts the challenge. The game of discovery goes on.

Fuel Cells

What a difference 50 years makes. Today, Americans drive more than any other nation, and consequently contribute a tremendous amount of carbon dioxide to the atmosphere. There are 148 million cars on the road in the United States, compared to 19 million in China and 9 million in India. But in 50 years, as those nations grow wealthier and more industrialized, more Chinese and Indian citizens will be driving cars. By 2050, Americans will be driving 233 million cars, the Chinese 514 million cars, and the Indians 610 million. How will our poor planet survive all that extra carbon dioxide? And where will we get the fuel to run all those vehicles, since petroleum is a finite resource?

Thankfully, a new technology—hydrogen fuel cells—is already in the works. In a hydrogen-powered vehicle, hydrogen and oxygen are mixed together to form an electrochemical reaction. One byproduct of this reaction is electricity, which runs the car. The other byproducts are water and heat, or steam. For decades, NASA has used hydrogen as a propellant for external combustion engines to thrust rockets into space. In fact, the reaction worked so well and burned so cleanly, that astronauts drank the pure water produced

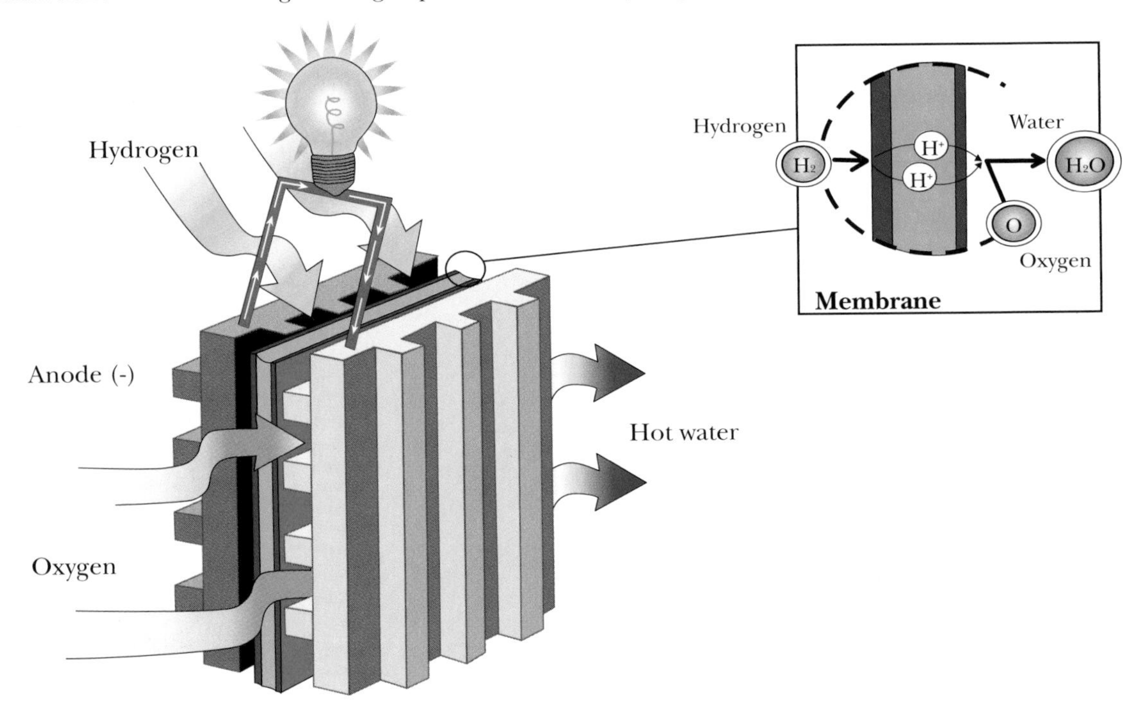

Above: Diagram of a hydrogen fuel cell shows hydrogen gas moving through the electrodes and combining with oxygen that has migrated through the opposite electrodes. This resultant oxidizing reaction causes a buildup of negative charge. The flow of electrons through the circuit is what creates electricity. Top left: The use of fuel composite buses is an important step in the reduction of automobile-caused pollution in metropolitan cities.

Toyota's "Fuel Cell Hybrid Vehicle" (FCHV), a prototype that is powered by a hydrogen fuel cell, was unveiled in 2001.

as a by-product of those engines. That is amazing, considering that inhaling the exhaust of a modern car is deadly.

The idea of a hydrogen vehicle is scientifically sound, though it is much too costly at the moment to produce mass-market vehicles. But all carmakers are experimenting with hydrogen vehicles; some already have test models on the road. Hydrogen is the most plentiful element in the universe, so we will always have lots of fuel.

HOW IT WORKS

Here is how the reaction works: First, the hydrogen fuel—in the form of two bonded hydrogen atoms (H_2)—is pumped into the fuel tank. When the hydrogen molecules are zapped with the fuel cell's catalyst, they split and lose their electrons. This reaction leaves the hydrogen molecules to conduct electricity and power the car. Next, the positively charged hydrogen protons pass through the catalyst, only to meet up with leftover electrons and incoming oxygen (O_2) molecules on the other side. The presence of the electrons sparks another reaction: The hydrogen protons reconnect with some electrons, and bond with the oxygen to form H_2O, or water. The heated water is expelled from the vehicle as steam.

Although this reaction works, there are significant hurdles to overcome before hydrogen fuel cell technology can be introduced in the marketplace. It is still too expensive to manufacture such engines for mass production. Scientists are still trying to perfect the fuel cell and the materials used to make it work. Although hydrogen is abundant in the universe, it is not easy to package as a fuel. A "pure" hydrogen fuel reaction is about 80 percent efficient, but in practice hydrogen fuel will probably have to be mixed with biofuels. That lowers the efficiency considerably.

THE CONVERSION ISSUE

Lastly, in the United States, converting the nationwide system of gasoline stations into hydrogen stations will be extremely costly. Nations like China and India have it easy by comparison, because they currently have little infrastructure and can more easily start from scratch.

Does this mean that hydrogen fuel cells are not good choices to run automobiles of the future? Not at all. It simply means that many more problems need to be solved before we are zipping down the highway in a hydrogen car. But the hard work will absolutely pay off. More humans will have a source of private transportation than at any other time in history, and the planet will not have to suffer the consequences.

Nanotechnology

In 1965 Gordon Moore, the co-founder of Intel, the world's largest manufacturer of computer processing units (CPUs), also known as computer "chips," predicted that the number of transistors packed onto a single chip would double annually. In 1975 he revised his estimate to every two years. Until very recently, the industry rule of thumb has been eighteen months. More transistors mean a centimeter-square chip can perform more functions for less money. But recently, engineers have warned that the so-called Moore's law will hit a wall sometime in the next two decades.

The reason is grounded in simple chemistry and physics. Transistors, the on-off switches that perform much of a chip's work, will become so small that it will be difficult to predict the course of an electric current. Already, the wires connecting the switches on a chip are 1/500th the width of a single human hair. To go much smaller, engineers will need new technology to scratch them onto a silicon wafer. Just how small can we build these important circuits? Soon we will reach the size of an atom!

Scientists regard this as a challenge and proof that we are entering the age of nanotechnology—machines, medicines, and products that are astoundingly small. The prefix "nano" means one-billionth of a unit of something. So a nanometer is one-billionth of a meter. How small is that? Consider: The diameter of a DNA molecule is two nanometers. The bond holding together two carbon atoms is even smaller: 0.15 (15/100) of a nanometer.

Above: An artist's rendering of a nanobot, a device that many believe will be the future of medical science. Scientists believe that nanobots could be used to break up dangerous clots and deposits in the human body. Top left: As computer chips get smaller, scientists must find ways to produce even smaller circuits.

BENEFITS AND DANGERS OF A NEW TECHNOLOGY

In the years to come, scientists and engineers will be building objects at the molecular level. Imagine, say, a camera so small that it could be injected into the human body, travel to a diseased

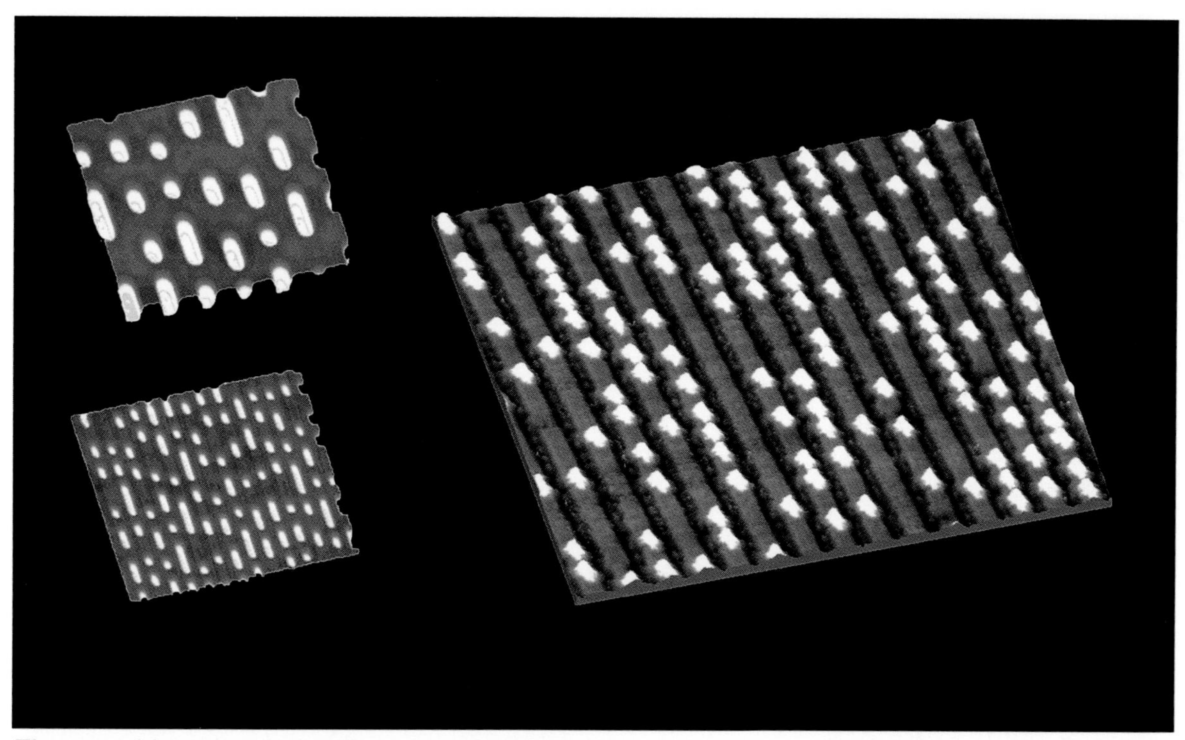

Three types of data storage media. The image on the right—taken by a scanning tunneling microscope and magnified 1,000 times more than the two images at left—shows individual atoms (in white) that are able to hold data. The image on the top left is of a CD-ROM, while the image beneath it is of a DVD.

site, and relay images back to doctors. Or consider a device that travels inside human vessels, dislodging and destroying fats that are clogging arteries. Envision paint capable of forming circuitry when it dries; paint a room, and it is instantly wired for electricity. These are just a few of the possible beneficial nano-inventions. Futurists have also warned that nations may invent ultra-small devices for surveillance and spying. They also fear that living things can be harmed by nanomachines that are unknowingly inhaled or absorbed into their skin. At such an early stage in this science, it is difficult to say how likely or unlikely such scenarios may be.

Right now, numerous universities, government agencies, and private companies are funding investigations into nanotechnology. Two recent inventions are still tackling the basic problem of building smaller computer chips. One scientist has developed an innovative way to "print" new circuits in seconds using an inkjet printer and special ink made of solvents and crystals of cadmium selenide. When the ink is applied to plastic, the solvent evaporates and the crystals remain. After a few passes through a special printer, the crystals build up and form working transistors. This approach would be a vast improvement over the current method of fabricating silicon chips, which takes place in large, sterile factories over periods as long as three weeks and often requires hundreds of steps.

Another scientist has built the world's smallest pen, capable of drawing complex circuitry on silicon chips. The "nanopen" uses a chemical called octadecanethial to draw wires about a millionth of an inch wide on gold-plated silicon. Chip manufacturers used to brag that they could etch wires on chips as thin as 180 nanometers. The nanopen draws lines just15 nanometers in width—enough to fit 80 million pages of information onto a square inch of silicon.

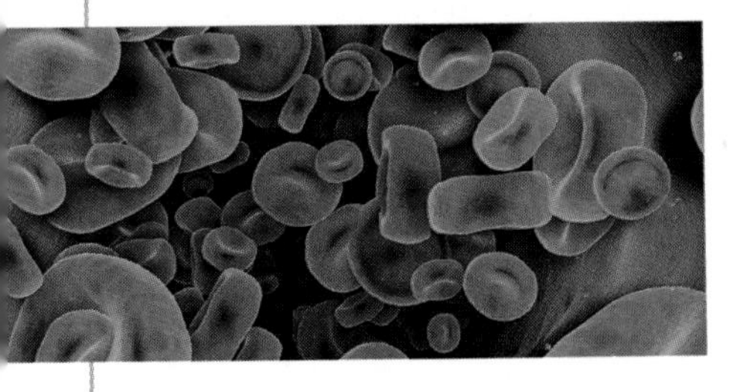

Proteomics

Proteins keep living organisms running. From enzymes like amylase, which break down starches in the body, to the hemoglobin that carries oxygen in the blood, the role of proteins in the key functions of the human body more than justifies their having a class of study all their own. Proteomics is the large-scale study of proteins, their structure, how they interact, and the functions they perform. And some of the most important aspects of proteomics are also linked to the rapidly evolving study of the genome. As mapping of the genome has been accomplished—although the current "map" is constantly being checked and rechecked—the

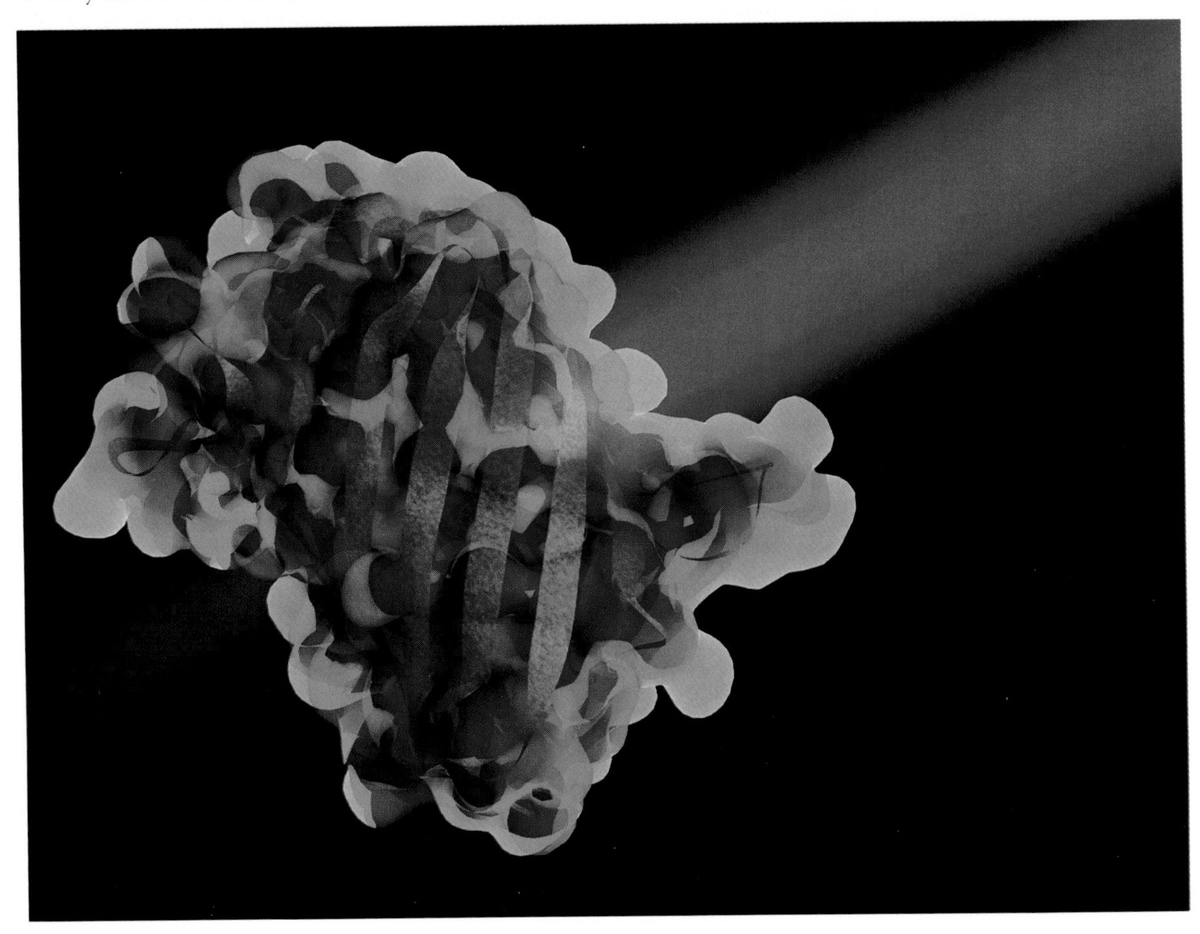

Above: Computer rendering of green fluorescent protein (GFP). Scientists are able to use GFP, which is found in the Pacific jellyfish and fluoresces green when blue light is shone on it, to track the movement of proteins or viruses within the body. Top left: Hemoglobin, one of the body's essential proteins, helps bond red blood cells to oxygen.

information it contains has given us a great deal of insight into the world of proteins and how they are coded. But there is still a long way to go.

RESEARCH FOR THE FUTURE

A proteome is the complete collection of an organism's proteins. The cataloging of all the proteins in the human body is a remarkable project that will be under way for quite some time. One of the more interesting aspects of this identifying and categorizing process is that there are considerably more proteins in the human proteome than there are genes encoded to create them. In 2003, the Human Genome Project announced that the coding of the human genome was, for the most part, finished. Since then, the estimated number of genes in the human genome has gone back and forth a bit, and is now put at between 30,000 and 40,000. However, at present no one can accurately estimate the number of proteins in humans, though scientists have made great strides in attempting to classify groups of proteins according to their general functions. In short, there is not only one protein for every gene. In fact, it is believed that the genes can code for a number of proteins at least 10 times that of their own. And since the number of genes in the genome is an estimate, the number of proteins in the proteome and how they work together is the next big thing on the genetics horizon.

The proper expression and functioning of proteins is necessary for the proper functioning of the human body. What makes a particular protein function—or, rather, not function—is at the very core of understanding a variety of diseases and difficulties that face the species. The human proteome is much larger than the human genome, and the information it contains is that much greater. So identifying and sequencing proteins is no small feat. First, hundreds of millions of copies of a cell must be made and their proteins extracted. Then those proteins are separated using a process called electrophoresis, after which they are broken down even further into their peptides. Finally, mass spectrometry is used to break down the peptides and analyze the masses of the individual parts, before all the information is fed into a database for further analysis.

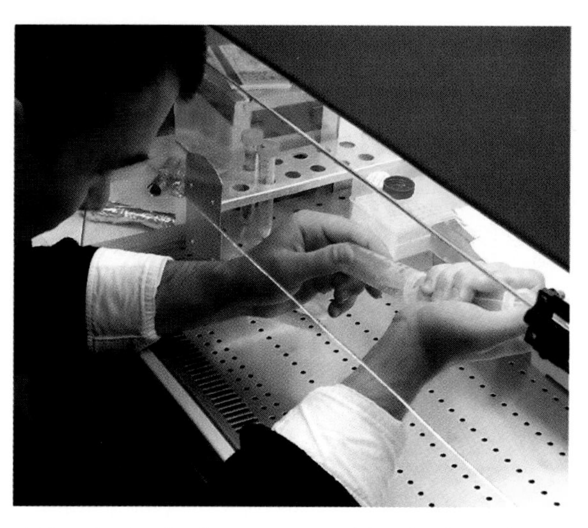

Although scientists have announced that they have, for the most part, mapped the human genome, there is still much to uncover about our biological universe.

CODING AND FOLDING

The coding of an individual protein by a gene is just the beginning. Genes can code for the creation of a particular protein, but then that protein has to "fold" over a certain way in order for it to function properly and be effective. The more that is understood about proteins, the better their ability to act as biomarkers, which can serve as indicators of the presence of disease or the efficacy of treatment. Understanding how proteins are made, how they function, and what prevents them from doing their job sets the stage for an entirely new era in the treatment of diseases. For example, scientists have found a protein that appears to guard the beta cells, the cells responsible for making life-sustaining insulin. To understand how this or any other protein functions, it is necessary for scientists to understand not only how proteins are coded, but what happens to proteins in the individual cells—how much of a particular protein an individual cell will synthesize, how those proteins are modified once they are created, and how they eventually interact with other proteins.

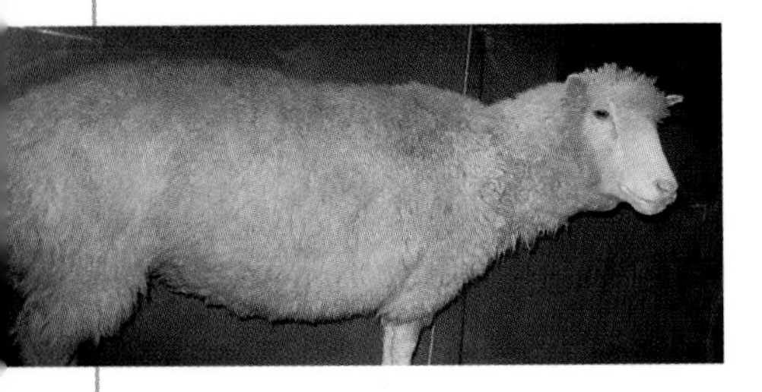

Genetic Engineering

One of the most exciting and highly debated areas at the forefront of scientific research in the twenty-first century is genetic engineering. The ability to grow disease-resistant food or the possibility of boosting an individual's defenses against a deadly disease through gene therapy are two examples of how genetic manipulation may change the way we live and thrive.

Genetic engineering, or genetic manipulation, is exactly what it sounds like: a means of engineering or changing the genetic information of a particular organism. Altering the genetic information of an organism can affect the development of proteins, how much of a particular protein is made, and how it is expressed. Introducing new DNA can result in new protein functions and changes in a variety of characteristics of an organism. There are plants that have been engineered to resist frost, for example. But genetic engineering, though it is now in the news more than ever, is nothing new. In the late 1970s, the first genetically engineered insulin was produced with a little help from bacteria and a technique that is still used today.

The gene for human insulin is extracted from healthy human cells and inserted into the plasmid of a bacterium. A plasmid is a ring of DNA found in bacteria that replicates on its own and does so outside of, and separate from, the chromosomal DNA. Once a bacterium has incorporated the insulin gene into its plasmid, it goes about its business: The plasmids replicate and the bacterium divides and reproduces, making insulin along the way and creating more insulin-producing bacteria as well. The insulin can then be collected and made available for diabetes patients everywhere. This form of insulin is now the norm and has replaced the need for diabetes patients to rely on purified pig and cow insulin.

Above: In recent years, scientists have been genetically modifying various crops, such as the grass shown here, in order to produce plants with greater nutritional and commercial benefits. Top left: The remains of Dolly the sheep, the first mammal to be successfully cloned from differentiated somatic cells, are on display in the Royal Museum of Scotland.

NEW COMBINATIONS

The production of insulin in bacteria is just one of thousands of developments that depend on recombinant DNA. Recombinant DNA is the "recombination" of genetic matter from one organism with the DNA of another. In the insulin example given here, the recombination takes place between the genetic material in the plasmid of a bacterium and the human insulin gene. Plasmids from bacteria and yeast are commonly used in this process. The plasmid ring of the bacterium or yeast is "cut" with an enzyme, and the same is done to a section of the chromosomal DNA from the second organism. The resulting piece of chromosomal DNA from the second organism will have unmatched base pairs at its end that will match the open ends of the plasmid DNA. Finally, ligase enzymes—enzymes that join ends of DNA—join the ends of the DNA from the two different organisms. The new, recombined plasmid DNA happily reproduces in the bacterium and the bacterium itself reproduces, creating clones that also contain the new DNA.

NEW USES

The future of gene therapy—the swapping out of defective or missing genes for ones that work properly—is thought to be a key to treating a variety of diseases, from diabetes to cystic fibrosis. Gene therapy goes hand in hand with proteomics. As the function and structure of crucial proteins are clearly identified and understood, this information can be used to design gene therapies that introduce the code for missing or malfunctioning proteins in patients. Recombinant DNA technology is currently used in everything from flu vaccines that keep us sniffle-free to growth hormones that keep cattle big and beefy. Genetically modified foods offer crops that are resistant to various pests and disease and that will grow faster. Genetically modified rice has been developed that is enriched with iron and other vitamins to help combat malnutrition. There is even a "banana vaccine" in the works—a banana that could protect people from contracting diseases like hepatitis B. But genetically modified food is a highly controversial topic. Those who actively oppose forging ahead are concerned there is no real way to predict the environmental and health risks or unwanted side effects that may result from these kinds of engineered products being introduced on a larger scale. Those in favor say the benefits will outweigh any risks. Both sides of this argument will have increasingly more fodder for discussion, as there appears no end in sight for the applications of recombinant DNA technology in industry and medicine.

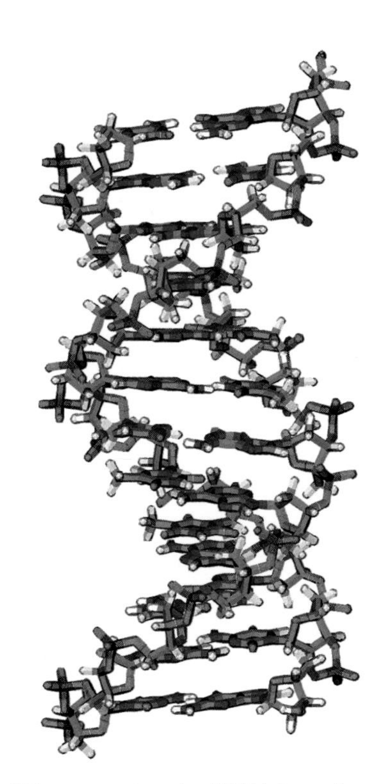

Telomeres—strands of DNA located at the ends of chromosomes—keep the DNA helix from unraveling and are linked with the aging process.

THE *REAL* FOUNTAIN OF YOUTH

Anyone hoping to live forever should study his or her telomeres. They are DNA strands that grow at the end of chromosomes to keep them from unraveling—not unlike the plastic or metal tips of shoelaces. Every time a cell divides, telomeres get shorter. When they get too short, the cell dies. Scientists think telomeres are linked to the aging process. Telomere studies are suggestive: obesity, smoking, and high stress all shorten telomeres—and one's life. Cancer cells are prodigious producers of telomerase, an enzyme that lengthens telomeres and makes cancer cells seemingly immortal. Could we live forever if we learn to extend our telomeres? That's the question everyone is dying to find out.

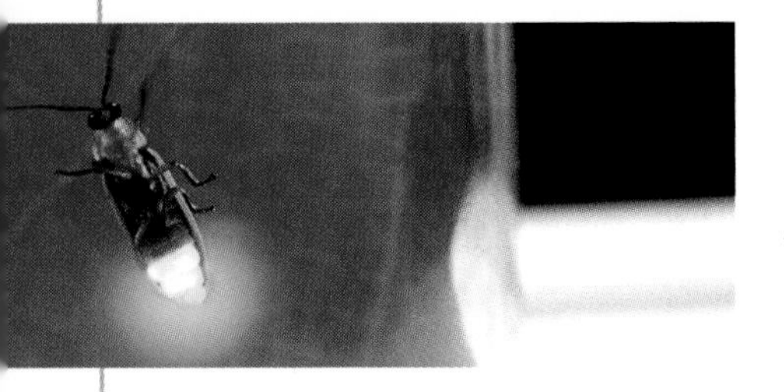

Advanced Materials

"I find out what the world needs, then I proceed to invent," said inventor Thomas Edison. The search for an effective new material can change the world. When Edison and his team of inventors were trying to perfect their version of the lightbulb, they auditioned thousands of materials to find the right filament, the wire that makes incandescent lightbulbs glow. They tried a variety of elements, among them platinum and tungsten, as well as the fibers of at least 6,000 different plants. The turning point came when Edison's team realized that they could use a carbonized version of ordinary sewing thread, as long as the bulb itself contained no air. They used a special pump to create a vacuum in the bulb. In an oxygen-free environment, their filament burned robustly.

Modern-day scientists and engineers are embarked on a similar quest for the ideal substance that will make their innovation come together. The branch of chemistry known as advanced materials science is concerned with discovering and developing new ceramics, composites, glasses, metals, and polymers that are useful to humans. Many companies and universities have advanced-materials labs, and the career choices within this field are dizzying, because a single material can be employed in many different ways. A new polymer, for instance, may be used in medicine, electronics, and aerospace engineering, and experts in each of these fields must carefully conceive, design, and test the polymer's new application in that realm. Advanced-materials chemists are also interested in how humans use knowledge and how they fail. If researchers were designing a new plastic film for storing data, for instance, they would want to know all the ways that humans store data, the typical mistakes they make while handling important information, as well as common problems with existing media. That is the only way the scientists can improve these products.

We have seen how scientists are developing new hydrogen fuel cells to power cars of the future. Another important requisite will be replacing heavy steel auto body frames with lightweight composites. These materials, used in skiing, snowboarding, fishing, and cycling, usually consist of carbon fibers bonded together to make tough, light compo-

Thomas Edison's improvements to the incandescent light bulb made it the first bulb that was economically viable for consumers. Top left: Often, the blueprints for scientific development are inspired by evolutionary developments found in nature.

nents. If we switched auto steel to composites, a car would weigh half what it does today and be just as strong or stronger. Many car parts are already made of composites, as are race cars and special aircraft. Cars that weigh less require less fuel to propel them. Fuel cells are only half the story; advanced materials are the other half.

INSPIRATION FROM LIVING THINGS

Scientists know that nature has had four billion years to perfect its designs. That is why, in the search for new materials, some scientists are drawing inspiration from living organisms. The classic example, seen earlier, is the design of photovoltaic panels (see chapter 5), which mimic photosynthesis performed in leaves. There are so many things we have yet to learn from nature. In the realm of genetics, we don't yet know how the folding of proteins impacts their use, and yet a brainless ribosome does. For the field of proteomics to succeed, we must become careful students of nature.

There are countless lessons waiting for us. Trees, insects, and mammals produce a wide variety of liquids, lubricants, and adhesives that might be worth borrowing from. Undersea creatures have shells of a geometrical and crystalline structure that is exceptionally durable, and may prove to be our tutors in engineering. Numerous creatures flit across the surface of the Earth and in the seas, using bioluminescence to light their way or distract predators. Their bodily wattage may hold the secret to lighting our homes, our businesses, and our world in the coming hyper-efficient age.

Motivating scientists is the understanding that there is no free lunch. Earth's resources will only become scarcer as the planet's population grows beyond 6.5 billion people. We must become better at using materials, and better at conserving them. In the European Union and Japan, governments are already mandating that makers of consumer electronics take back and reprocess computers, cables, printers, cell phones, and other forms of electronic trash that would only end up in landfills and pollute the planet with their toxins. Such laws have proved costly for these companies, who are now engaged in an all-out race to build an environmentally friendly computer. And we all benefit from this new paradigm: the companies that market the first safe, recyclable products, the scientists who develop them, and the rest of us, who are privileged to live in a world made better by advanced materials.

Scientists have studied the unique water-retention system employed by this Namibian beetle in order to develop plastics that will be useful to inhabitants of water-deprived areas.

WATER-CONSERVING MATERIAL—FROM A BUG!

The *Stenocara gracilipes* beetle that lives in Africa's Namib Desert has an unusual method for harvesting water. Scientists noticed that the beetle's bumpy shell is coated with waxy lipids that render the smooth part of the shell hydrophobic, or water-resistant. The nonwaxy bumps are hydrophilic. When foggy clouds blow across the desert, the beetle tilts its body to intercept the wind. Water vapor condenses to liquid on the creature's shell. When a droplet reaches 5 millimeters in size, it slowly trickles down the beetle's back to its mouth. Scientists are employing the same water-capturing microstructures to create plastics that will help peoples in parched parts of the planet collect water from the atmosphere. The biomimetic design could help us fabricate better roofing materials, distilling equipment, and dehumidifiers.

Drug Design

Like a dressmaker designing patterns at an upscale couturier, or an architect poring over blueprints for a custom-made home, scientists are creating tailor-made drugs to tackle specific diseases, based on an increasing knowledge about how chemicals are structured and how they fit together. As scientists gain insight into how a particular disease functions—because of a malfunctioning protein, for example—they can more easily design drugs that target the problem. By combining the benefits of proteomics, genetic engineering, and advanced computer technology, the process of designing drugs is becoming more efficient and more precise.

A variety of techniques and advances are driving the drug design industry. Combinatorial chemistry uses various molecular compounds as chemical "building block" units, systematically combining and rearranging them to create a huge collection of new chemicals, each one varying a bit from the others. These can then be screened en masse, and those candidates that show the most promise in treating the targeted protein can then be synthesized for further testing. This process—which may also employ the use of a robotic combinatorial system—allows scientists to create and then screen thousands of chemicals at once.

The structure of a drug is as important as the chemicals that make it up. A large part of drug design is described as structure-based design and focuses on the structure of the target molecule and the drug being designed for it. Techniques like X-ray crystallography are used to identify the structure of proteins, making it possible for scientists to structure a drug that will best target the molecule in question. Much like with the enzyme-substrate model, researchers need to find the right "fit" in order for a drug to be effective. For this reason, computer simulation and modeling now play a major role in drug design. Modeling can be used to analyze the molecular interactions that will occur between drug and target in order to predict how drugs will function and how well they will be able to bind to their target site. And all of these techniques work together. In AIDS research, for example, scientists discovered that the survival of HIV (human immunodeficiency virus) was

Above: Scientists often study the chemical properties of plants in order to develop effective drugs. Natural remedies are still used by many people around the world as an alternative to pharmaceuticals. Top left: Aqueous solutions containing phenol, such as the ones seen here, were used as antiseptics in the nineteenth century.

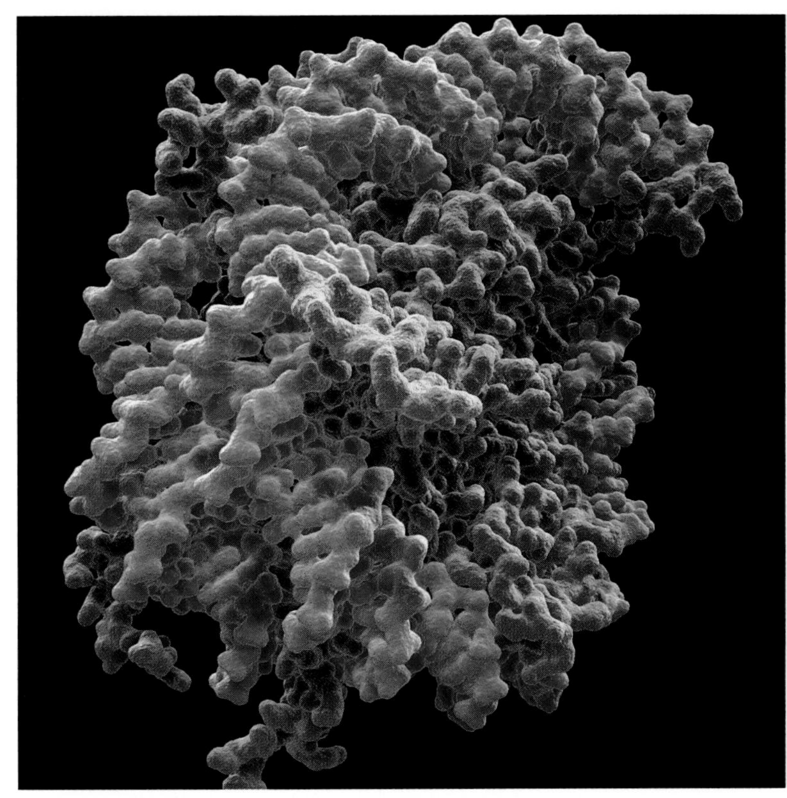

Above: This model of a nucleosome was drawn using data collected by X-ray crystallography. Here, the protein (orange) is covered by the chromosomal DNA (green). Below: As scientists learn more about the diseases that attack our bodies, they are able to design more effective pharmaceuticals.

linked to the HIV protease, a viral enzyme. Using X-ray crystallography, scientists could see the structure of the HIV protease, and then, using computer-based modeling, they could spin it around and examine it from every angle, analyzing its properties and narrowing down the list of possible molecules—perhaps generated through combinatorial chemistry—that might be able to block its activities.

KEEPING UP

Science has always gotten some of its best ideas from nature, and this tendency is showing no sign of changing. Many pharmaceutical companies are examining properties in natural plants and then using molecular modeling to reconstruct the compounds in a laboratory setting. But science is also often faced with the daunting task of keeping up with nature, and drug design is relied upon more and more to help researchers do just that. The more that is understood about the human genome and proteome, the more real the possibility that drugs will be designed to work best for an individual's specific genetic makeup. But one of the challenges facing scientists is that nature keeps things lively—and changeable. Many reliable drugs that have stood the test of time find themselves facing evolutionary challenges. For example: Penicillin, the mold-derived antibiotic discovered in 1928, is probably the most widely prescribed antibiotic today. However, its efficacy in fighting bacteria has been challenged, as bacteria have now developed their own defense against their arch-nemesis: an enzyme that destroys it. The same natural wonders that have inspired and shaped scientific discovery throughout time are also responsible for throwing up roadblocks, which may then send research in new, powerful directions. We are constantly witness to the uneasy, unpredictable relationship—at times mutually beneficial, at other times combative—between the world around us and the science that studies it.

GLOSSARY

A

ABSOLUTE TEMPERATURE AND ABSOLUTE ZERO The temperature scale with zero as lowest possible temperature, also known as the Kelvin scale. Zero degrees Kelvin equals -273 degrees Celsius. At absolute zero, particles are void of all heat and atomic movement has all but ceased.

ACID Having a pH less than 7.

ACID RAIN (ACID PRECIPITATION) Precipitation containing pollutants, rendering it acidic.

ACTIVATION ENERGY Minimum energy needed to start a reaction.

AEROBIC Requiring or using oxygen.

ALKALINE Having a pH greater than 7. Basic.

ALLOTROPE Different forms of a particular element. Example: Graphite and diamonds are allotropes of carbon.

ALLOY A new metal made from two or more metallic elements.

ALPHA DECAY Radioactive decay of the nucleus which emits an alpha particle.

ALPHA PARTICLE Composed of two neutrons and two protons and emitted from the nucleus of a radioactive atom during decay.

AMINO ACID A basic building block of protein.

ANABOLISM Constructive aspect of metabolic process in which more complex substances are formed from simpler ones.

ANAEROBIC Occurring in the absence of oxygen.

ANION A negative ion.

ANODE The positive electrode of a battery or electrochemical cell.

ATMOSPHERE The air surrounding Earth, made up of specific layers.

ATOM The smallest component of an element that still has all the same properties of that element.

ATOMIC MASS NUMBER The total of protons and neutrons in an atom.

ATOMIC NUMBER The number of protons in a particular atom, which identifies the element.

B

BACTERIA One of the three domains of life. Single-celled organisms without nuclei.

BASE Having a pH greater than 7.

BECQUEREL Unit of radiation, with 37 billion Bq equal to 1 curie.

BETA DECAY Radioactive decay of the nucleus, which emits a beta particle.

BETA PARTICLE An electron or a positron, emitted during radioactive decay.

BIG BANG A cosmology theory that assumes the universe began as a gigantic explosion.

BIOSPHERE All the areas of Earth's crust, water, and atmosphere where life exists.

BUFFER A compound that resists changes to its pH.

C

CALORIE Amount of energy needed to raise the temperature of one gram of water one degree Celsius.

CARBOHYDRATE Consisting of carbon, hydrogen, and oxygen; includes sugars, starches, and cellulose.

CARCINOGEN Cancer-causing substance.

CATABOLISM Destructive aspect of metabolic process in which more complex substances are broken down into simpler ones.

CATALYST Substance used to increase the rate of a chemical reaction but is not part of the reaction.

CATHODE The negative electrode of a battery or electrochemical cell.

CATION A positive ion.

CELLULOSE Carbohydrate that forms part of the cell walls of plants.

CELSIUS A temperature scale that gives zero as the freezing point of water and 100 as its boiling point.

CHEMICAL ELEMENT An element is composed of only one type of atom.

CHLOROFLUOROCARBONS A greenhouse gas containing carbon, chlorine, fluorine, and sometimes hydrogen.

CHLOROPHYLL A pigment, usually green, used to capture light energy for photosynthesis.

COMBUSTION The act of burning a substance and releasing heat.

COMPOUND Substance made up of two or more elements.

CONDENSATION Process by which a gaseous substance changes into to a liquid state.

CONDUCTION Transfer of heat by means of contact between neighboring molecules.

CONVECTION Transfer of heat from one place to another by actual motion of the hot material.

COSMIC BACKGROUND RADIATION Thermal blackbody radiation that permeates the universe; it has a temperature of 2.7 degrees Kelvin and is left over from the big bang.

CURIE Unit of radiation named for scientist Marie Curie and equal to 37 billion Bq.

D

DEOXYRIBONUCLEIC ACID (DNA) Material found in the cell nucleus encoded with genetic information.

DEUTERIUM An isotope of hydrogen sometimes referred to as "heavy" hydrogen, which has one proton and one neutron.

DISACCHARIDE A sugar composed of two monosaccharides or simple sugars.

E

ELECTROMAGNETISM The study of the combined effects of electricity and magnetism.

ELECTRON Negatively charged particle that orbits the nucleus of an atom.

ELECTRON CONFIGURATION Arrangement of electrons in orbitals around the nucleus of an atom.

ELEMENT See *chemical element.*

ELEMENTARY PARTICLE A basic particle out of which all other particles, or matter, is made.

EMULSIFIER A substance that helps the suspension of one liquid in another.

ENDOTHERMIC A reaction that absorbs energy.

ENERGY The capacity or ability to do work. Types of energy include chemical, mechanical, thermal, and electrical.

ENZYME A protein that acts as a catalyst in biochemical reactions.

ESSENTIAL AMINO ACID An amino acid that is not made by living things and therefore must be supplied through their diet.

EXOTHERMIC A reaction that gives off energy.

F

FAHRENHEIT A temperature scale that gives 32 as the freezing point of water and 212 as its boiling point.

FATS Energy source made up of fatty acids and glycerol. See *lipid.*

FATTY ACID Organic molecule that is a major component of fats.

FIRST LAW OF THERMODYNAMICS The total amount of energy in the universe is constant and cannot be created or destroyed, only converted from one form to another.

FOSSIL FUEL Fuel derived from long-dead organisms; includes oil, coal, and natural gas.

FUEL CELL A device that uses chemical reactions to produce electricity.

FUEL ROD A metal tube containing fuel to be used in a nuclear reactor.

FULLERENE An allotrope of carbon discovered in 1985.

G

GAMMA DECAY Radioactive decay of the nucleus, which emits gamma particles or gamma rays.

GAMMA RAY Electromagnetic radiation, invisible to the human eye, of the highest energy and frequency.

GAMMA PARTICLE A photon or light particle.

GREENHOUSE EFFECT The trapping of heat within the Earth's atmosphere, caused by greenhouse gases.

H

HALF-LIFE Time required for one-half of the atoms in a radioactive sample to disintegrate.

HEAT Also called thermal energy, heat moves between substances of varying temperatures.

HYDROCARBON An organic molecule or compound consisting of carbon and hydrogen.

HYPOTHESIS A scientist's best guess or theory that has yet to be proved.

I

INHIBITOR Substance used to block or decrease the rate of a chemical reaction.

INTERNAL ENERGY The total energy, kinetic and potential, of a particular system.

ION A particle with a positive or negative charge.

ISOTOPE One of several forms of a chemical element possessing the same number of protons but a different number of neutrons and different atomic weight.

K

KELVIN SCALE See *absolute temperature scale.*

KINETIC ENERGY The energy of motion.

L

LAW OF CONSERVATION OF ENERGY See F*irst Law of Thermodynamics.*

LIPID Type of carbohydrate that includes fats, oils, steroids and waxes.

LOCK-AND-KEY MODEL Theory explaining how enzymes and their specific substrates work together and bind.

M

MASS NUMBER See *atomic mass number.*

MATTER All things that have mass and take up space.

METABOLISM Sum of catabolic and anabolic activities in an organism.

METAL A type of element consisting of metallic bonds; good conductors of electricity.

METALLIC BOND Bond found in metals in which electrons roam freely between the bonded atoms.

MOLECULE A combination of atoms, bound together by an electromagnetic force, that forms the smallest unit of a compound.

MONOMER An individual unit or molecule that makes up a polymer.

MONOSACCHARIDE Simple sugar, such as glucose.

N

NANOMETER One-billionth of a meter.

NANOTECHNOLOGY The design, study and building of microscopic devices that are 100 nanometers or less.

NEUTRALIZATION Reducing the acidity or alkalinity of a substance by adding acids or bases as needed.

NEUTRON The neutral particle of the atomic nucleus.

NONMETAL Type of element that readily gains electrons from other atoms.

NONRENEWABLE RESOURCE A resource, such as fossil fuels, that cannot be readily renewed because it takes so long to form.

NUCLEAR FISSION The breaking apart or splitting of nuclei into two or more smaller parts.

NUCLEAR FUSION The combining of smaller nuclei to make larger, heavier nuclei.

NUCLEIC ACID Molecular substances containing genetic information, such as DNA and RNA.

NUCLEOTIDE A basic building block of nucleic acids.

NUCLEUS The center of the atom, containing protons and neutrons. Also refers to the center of a living cell, containing genetic material.

O

ORBITAL Area within an atom in which it is most likely that electrons will be found.

ORGANIC CHEMISTRY A division of chemistry dedicated to the study of organic compounds, including their structure, properties, reactions, and synthesis.

ORGANIC COMPOUND A compound that contains carbon and usually hydrogen, often including oxygen or nitrogen as well.

ORGANISM Any living thing, including bacteria, plant life, animals, and humans.

P

PEPTIDE Two or more amino acids bonded together. The bond is called a peptide bond.

PERIODIC TABLE Table of all known chemical elements organized according to their properties, including their atomic numbers.

PH Stands for "potential of hydrogen." A measure of the acidity or alkalinity of a substance. The lower the pH, the more acidic a substance is. Neutral pH is 7.

PHOTOSYNTHESIS In a plant, the creation of sugar from carbon dioxide and water in the presence of sunlight and chlorophyll.

POLYMER A large molecule made up of many smaller, repeated components (monomers). Starch is a polymer made up of many individual sugar molecules.

POLYPEPTIDE Many amino acids joined together by peptide bonds. Example: protein.

POLYSACCHARIDE Many monosaccharides (simple sugars) joined together. Example: cellulose.

POTENTIAL ENERGY The energy of an object at rest.

PRODUCTS That which is produced in a chemical reaction.

PROTEIN Molecule composed of chains of amino acids, responsible for performing many of the body's important functions.

PROTON Particle in the nucleus of an atom, containing a positive charge.

Q

QUANTUM A discrete amount or quantity.

QUANTUM THEORY The theory that energy comes in small units, or quanta (also called quantum mechanics).

QUARK Subatomic particles that make up protons and neutrons.

R

RADIATION The emission of energy in the form of waves. Also, particles emitted by radioactive substances.

RADIOACTIVITY The disintegration of unstable atomic nuclei with the emission of radiation or particles.

REACTANTS The molecules or substances that enter into a chemical reaction.

RESIN IDENTIFICATION NUMBER Identifying number on plastic goods indicating their potential to be recycled.

RIBONUCLEIC ACID (RNA) Material found in the cell nucleus, used in the replication of DNA and the coding of new proteins.

S

SACCHARIDE Group of carbohydrates containing sugars.

SOLUTE Substance that can be dissolved in a solvent to form a solution.

SOLUTION Resulting mixture after a solute has been dissolved in a solvent.

SOLVENT Substance that can dissolve a solute to form a solution.

SPECIFIC HEAT The amount of energy needed to raise one gram of a specific substance one degree Celsius, measured in calories.

SPECTRUM The array of electromagnetic radiations spread out in order of wavelength.

STRATOSPHERE Section of Earth's atmosphere existing between 7 and 30 miles above the Earth's surface.

SUBATOMIC PARTICLES Particles that make up an atom, including protons, neutrons, and electrons.

SUBSTRATE Generally, a substance that is acted upon. Substrates are substances acted upon by enzymes, forming the enzyme-substrate complex. A substrate can also be the top layer of a silicon chip that is etched to form a circuit.

SURFACE TENSION The result of the attraction between water molecules as seen on the surface of the liquid.

SURFACTANT Also called a surface-active agent, it can reduce the surface tension of a liquid.

SUSPENSION Small particles dispersed, but not dissolved, in a liquid.

T

THERMOCHEMISTRY The study of the heat or energy changes that take place during chemical reactions.

THERMOPLASTICS Plastics that can be heated and cooled a number of times without losing their properties.

THERMOSETTING PLASTICS A polymer material, such as epoxy, that must be mixed with another substance in order to be activated and cannot be reformed once it has "cured" or dried to its final form (unlike thermoplastics).

TRIGLYCERIDE A form of fat consisting of three fatty acids and one glycerol.

TROPOSPHERE Section of Earth's atmosphere existing immediately above Earth's surface and extending 5 to 9 miles above it.

V

VALENCE ELECTRON Electron located in the outermost shell of an atom.

VAPOR Gaseous form of a substance that is normally a liquid or solid under ordinary conditions.

VOLTAIC CELL Named for Alessandro Volta, an electrochemical cell that depends on spontaneous chemical reactions to produce electricity.

W

WAVELENGTH The distance between equivalent points of a wave.

WORK A transfer of energy from one object to another that results in the movement or displacement of the object that is being acted upon.

X

X-RAY Radiation that falls between gamma and ultraviolet rays on the electromagnetic spectrum.

FURTHER READING

BOOKS

Alpher, Ralph, and Robert Herman. *Genesis of the Big Bang*. New York: Oxford University Press, 2001.

Cobb, Cathy, Monty L. Fetterolf, and Jack G. Goldsmith. *Crime Scene Chemistry for the Armchair Sleuth*. Amherst, NY: Prometheus Books, 2007.

Cobb, Cathy, and Harold Goldwhite. *Creations of Fire: Chemistry's Lively History from Alchemy to the Atomic Age*. New York: Perseus Books, 2002.

Curie, Eve, and Vincent Sheean. *Madame Curie: A Biography*. New York: Da Capo Press, 2001.

Emsley, John. *Nature's Building Blocks: An A–Z Guide to the Elements*. New York: Oxford University Press, 2003.

———. *The 13th Element: The Sordid Tale of Murder, Fire, and Phosphorus*. Hoboken, NJ: Wiley, 2000.

Faraday, Michael. *The Chemical History of a Candle*. Mineola, NY: Dover Publications, 2003.

Finlay, Victoria. *Color: A Natural History of the Palette*. New York: Ballantine, 2003.

Ford, Leonard A. *Chemical Magic*. Mineola, NY: Dover Publications, 1993.

Garfield, Simon. *Mauve: How One Man Invented a Color That Changed the World*. New York: W. W. Norton, 2001.

Hager, Tom. *Linus Pauling: And the Chemistry of Life*. New York: Oxford University Press, 2000.

Houk, Clifford C., and Richard Post. *Chemistry: Concepts and Problems: A Self-Teaching Guide*. Hoboken, NJ: Wiley, 1996.

Jonnes, Jill. *Empires of Light: Edison, Tesla, Westinghouse, and the Race to Electrify the World*. New York: Random House, 2003.

Karukstis, Kerry K., and Gerald R. Van Hecke. *Chemistry Connections: The Chemical Basis of Everyday Phenomena*. San Diego: Academic Press, 2000.

Maddox, Brenda. *Rosalind Franklin: The Dark Lady of DNA*. New York: HarperCollins, 2002.

Mascetta, Joseph A. *Chemistry the Easy Way*. Hauppauge, NY: Barron's Educational Series, 2003

Pauling, Linus. *General Chemistry*. New York: W. H. Freeman, 1970.

Schwarcz, Joe. *The Genie in the Bottle: 67 All-New Commentaries on the Fascinating Chemistry of Everyday Life*. New York: Owl Books, 2002.

Seifer, Marc J. *Wizard: The Life and Times of Nikola Tesla: Biography of a Genius*. New York: Citadel Press, 2001.

Simon, Jonathan. *Chemistry, Pharmacy and Revolution in France, 1777–1809*. Burlington, VT: Ashgate Publishing, 2005.

Stocker, Jack H. *Chemistry and Science Fiction*. Washington, DC: American Chemical Society, 1998.

Tesla, Nikola. *My Inventions: The Autobiography of Nikola Tesla*. Hart Brothers Publishing, 1982.

Watson, James D. *The Double Helix: A Personal Account of the Discovery of the Structure of DNA*. New York: Touchstone, 2001.

———. *Genes, Girls, and Gamow*. New York: Oxford University Press, 2001.

WEB SITES

American Chemical Society | www.acs.org

American Physical Society | www.aps.org

Chemical Heritage Foundation | www.chemheritage.org

Environmental Protection Agency | www.epa.gov

NASA | www.nasa.gov

National Human Genome Research Institute | www.genome.gov

Nobel Prize | www.nobelprize.org

Royal Society of Chemistry | www.rsc.org/chemsoc/

Science Daily | www.sciencedaily.com

Smithsonian Institution | www.smithsonian.edu

United States Geological Survey | www.usgs.gov

AT THE SMITHSONIAN

James Smithson, a well-regarded chemist and mineralogist, bequeathed the bulk of his estate "to found at Washington, under the name of the Smithsonian Institution, an Establishment for the increase and diffusion of knowledge." Today it is the world's largest museum complex and research organization. Along with its affiliate museums, associated laboratories, and extensive archives, the Smithsonian has long been a resource for scientists and those who wish to expand their understanding about various scientific disciplines, their history and their future.

Among its vast holdings are a number of artifacts pertaining to the history of the chemical sciences. The National Museum of American History holds a vast share of these objects, while the National Museum of Natural History conducts research specifically in the Department of Mineral Science, where the chemical and mineralogical analysis of meteorites and geochemistry of metamorphic rocks and fluids takes place.

The two museums are located adjacent to each other on the National Mall between 10th and 14th Streets and Constitution Avenue NW in Washington, D.C. The research and collections at both museums offer numerous examples of how chemistry has shaped our existence over time.

Above: The National Museum of American History in Washington, D.C. Top right: The Nier Mass Spectrograph was built in 1940 and represented a turning point for American scientists' understanding of nuclear fission.

ENERGY & POWER COLLECTION

From the tiniest atom to the largest turbine, the museum's collection on energy and power sheds light on how we have harnessed the means to make our civilization hum. Featured artifacts in this collection include water turbines, windmills, and steam, gas, and diesel engines. There are also examples of equipment used for oil exploration and coal mining, and even bubble chambers, which physicists use to study subatomic particles.

The highlight of the energy and power collection is its wide range of artifacts that depict the history of electrical power: generators, batteries, cables, transformers, and even early photovoltaic cells. An especially captivating display is a group of Thomas Edison's earliest lightbulbs. Other selected artifacts include Alessandro Volta's voltaic pile, the first fluorescent lamps made in the United States by General Electric, the original turbines from Niagara Falls, one of the first direct-current generators, and the Nier Mass Spectrograph, which was instrumental in the early days of nuclear fission.

HEALTH & MEDICINE COLLECTION

This collection contains artifacts once used or collected by some of the scientists mentioned in this book. For example, the collection houses early X-ray apparatuses, including one of Wilhelm Röntgen's tubes. It also features a penicillin mold that was harvested by Alexander Fleming's history-changing experiments, and a sample of Jonas Salk's original polio vaccine. The collection is not limited to early medical history, however, and has recently acquired the first artificial heart implanted in a human, the earliest genetically engineered drugs, and additional materials related to David, the "Bubble Boy." The collection also includes artifacts as varied as artificial limbs and CT scanners to the contents of a medieval apothecary shop and the latest alternative medicines.

INDUSTRY & MANUFACTURING COLLECTION

This collection documents the role of industry in American life. The history of the first synthetic polymer, nylon, is brought to life as one looks upon the nation's original nylon-manufacturing machinery. The birth of the silicon chip, the brain of the modern computer, is also highlighted. For a change of pace, the collection also offers visitors the opportunity to view more than 460 episodes of the television series Industry on Parade, which celebrated American industry in the 1950s. Another highlight of this collection is the Bakelizer, a lab tool used by chemist-inventor Leo H. Baekeland. The Bakelizer produced vast quantities of the first totally synthetic plastic, Bakelite, which was used in the colorful costume jewelry of the era.

MEASURING & MAPPING COLLECTION

The measuring and mapping collection handles Thomas Jefferson's thermometer, a pocket compass used by Meriwether Lewis and William Clark on their expedition across the American West, and a timing device once part of the pioneering motion studies conducted by Eadweard Muybridge in the late 1800s. Of particular interest is an 1877 astronomical spectroscope, originally designed to be used in conjunction with a telescope in order to study the light of the Sun.

SCIENCE & MATHEMATICS COLLECTION

Instruments in this collection range from early American telescopes to lasers. Rare glassware and other artifacts from the laboratory of Joseph Priestley, the discoverer of oxygen, are among the scientific treasures here. Special objects featured in this collection include the Beckman DU spectrophotometer; the Ferrel Tide Predictor, inspired by a machine once developed by William Thomson (Lord Kelvin) to study the periodic motions that produce tides; and Hartman's Planispheric Astrolabe, an astronomical calculating device used from ancient times into the eighteenth century. The famous discovery of DNA is brought to life by a piece of Watson and Crick's original model of the double helix molecule: one of four brass templates illustrating the base pairings of adenine and thymine or cytosine and guanine.

If a visit to Washington, D.C., is not on the agenda, the Smithsonian offers extensive online and educational resources at: http://www.si.edu/science_and_technology/

Right: The Ferrel Tide Predictor was used by the U.S. Coast and Geodetic Survey from 1883 to 1910. The accuracy of this device led many to consider the possibilities of machine-derived computation. Left: Leo Hendrik Baekeland's original Bakelizer, which produced the synthetic material Bakelite by placing phenol and formaldehyde under pressure at high temperatures. Top: The first electro-hydraulic heart implanted into a human, the AbiCor Total Artificial Heart is powered by batteries and is completely concealed within the body of the patient.

INDEX

D

E

F

O

P

U

V

W

X

Y

ACKNOWLEDGMENTS & PICTURE CREDITS

The publisher wishes to thank consultant Eugene Jarosewich, National Museum of Natural History; Ellen Nanney, Senior Brand Manager with Smithsonian Business Ventures; Katie Mann and Carolyn Gleason with Smithsonian Business Ventures; Collins Reference executive editor Donna Sanzone, editor Lisa Hacken, and editorial assistant Stephanie Meyers; Hydra Publishing president Sean Moore, publishing director Karen Prince; editor Myrsini Stephanides, consultants, James M. Carothers, Ph.D. and Jeffrey P. Filippini, MA; copy editor Glenn Novak, editorial director Aaron Murray, art director Brian MacMullen, designers Erika Lubowicki, Ken Crossland, Eunho Lee, Pleum Chenaphun, Lee Bartow, and Gus Yoo, editors Marcel Brousseau, Michael Smith, Amber Rose, Rachael Lanicci, Suzanne Lander, Lori Baird, and John Finkbeiner; picture editors Sylke Jackson and Liz Mechem; picture researcher Ben DeWalt; picture researcher Joan Mathys; indexer Jessie Shiers; Wendy Glassmire of the National Geographic Society; Harriet Mendlowitz of Photo Researchers, Inc.

PICTURE CREDITS
The following abbreviations are used: PR Photo Researchers, Inc.; SPL Science Photo Library; JI © 2006 Jupiterimages Corporation; SS ShutterStock; IO Index Open; IS ©iStockphoto.com; NOAA–National Oceanic and Atmospheric Association; OAR–Oceanic and Atmospheric Research; NURP–National Undersea Research Program; USGS–United States Geologic Survey; NSF–National Science Foundation; NASA–National Aeronautics and Space Administration; NLM–Courtesy of the National Library of Medicine; SI–Smithsonian Institution; SIL–Smithsonian Institute Library; AP–Associated Press; LoC–Library of Congress; NGIC–National Geographic Image Collection; CDC–Center for Disease Control (t=top; b=bottom; l=left; r=right; c=center)

Introduction
IV JI **V** JI **VI** SPL/ Maximilian Stock **1t** IS/Christian Anthony **1b** JI **2** Clipart **3tl** PR/David R. Frazier **3br** SS/PhotoCreate

Chapter 1: Atoms and Elements
4 SS/Rebecca Dickerson **5t** JI **5b** SS/Arlene Jean Gee **6tl** JI **6** LOC **7tl** LOC **7bl** LOC **7br** LOC **8tl** NIST/Joseph Stroscio & Robert Celotta **8br** SPL/Lawrence Berkeley Laboratory **9tr** PR/Professor Peter Fowler **9br** LOC **10tl** SS/Chris Harvey **10b** LOC **11tl** LOC **11br** LOC **12tl** SS/George Unger IV **12bl** LOC **13t** SS/George Unger IV **13b** JI **14tl** PR/E.R. Degginer **14bl** NIST/D. Bright & D. Newbury **15** NOAA/Y. Berard, Oceanic Museum of Monaco **16tl** SS/Andraz Cerar **16bl** PR/Russell Lappa **16r** SS/Steffen Foerster **17tl** NIST **17br** SS/Roman Milert **18tl** JI **18r** SS/Jubal Harshaw **19tl** LOC **19bl** PR **19tr** IO

Chapter 2: Compounds
20 JI **21t** LOC **21b** IS/Matteo Natale **22tl** SS/ Sierpniowka **22br** SIL **23b** CDC/Minnesota Department of Health, R.N. Barr Library; Librarians Melissa Rethlefsen and Marie Jones **23r** LOC **24tl** SS/Coko **24bl** SS/Jay Crihfield **24br** LOC **25l** PR/Eye of Science **25tr** LOC **26tl** JI **26br** CDC **27bl** CDC **27tr** LOC **28tl** JI **29t** IS/Matteo Natale **29bl** SS/Radu Razvan **29br** SS/Ed Isaacs **30tl** SS/GSK **30bl** AP/Shari Lewis **30r** LLNL **31tr** SPL/Kenneth Eward/Biografx **31br** LOC **32tl** SS/Chris Rabkin **32tc** LOC **32bl** LOC/Doris Ulmann **33t** PR/Andrew Lambert **33bl** JI **34tl** SS/Elke Dennis **34c** SS/Stephen Beaumont **34r** SS/Elena Kalistratova **35t** SS/Jeremy Feng **35c** USDA/Jonathan House

Chapter 3: Matter In All Its Forms
36 SS/Doug Baines **37t** NOAA **37b** NOAA/Mr. Ardo X. Meyer **38tl** SS/Paul Cowan **38bl** PR/Alfred Pasieka **38r** SS/Dan Bannister **39tl** SS/Zimmytws **39br** JI **40tl** LOC **40bl** SIL **40r** NOAA Photo Library, NOAA Central Library; OAR/ERL/National Severe Storms Laboratory (NSSL) **41bl** JI **41tr** LOC **41cr** LOC **42tl** SS/JJJ **42bl** SS/Glen Jones **42r** SS/Ronen **43tr** NASA/J.P. Harrington & K.J. Borkowski, University of Maryland **43cr** NOAA/Dr. Yohsuke Kamide, Nagoya University **44tl** IO/FogStock, LLC **44r** IO/Photolibrary.com PTY. LTD. **45tl** SS/Jaime Wilson **45tr** SS/Peter Clark **45br** NOAA/NURP/OAR-Alaska Department of Fish and Game **46tl** SS/Losevsky Pavel **46r** SS/TAOLMOR **47tl** IS/Jane Norton **47bl** JI **47br** CDC/Bob Sanders **48tl** JI **48bl** NOAA/Grant W. Goodge **48r** SS/Dan Collier **49tl** SS/Andraz Cerar

Chapter 4: Chemical Changes
50 PR/Dr. Arthur Winfree **51t** SS/Jason Hassig **51b** JI **52tl** SS/Tim Hope **52r** SS/Jack Dagley Photography **53tr** SPL/Charles D. Winters **54tl** SS/John C. Hooten **54b** SS/Nadejda Ivanova **55t** IS/Loic Bernard **55bl** CDC **56tl** IS/Johann Helgason **56b** SS/Jurgen Ziewe **57** SS/Kim French **58tl** JI **58bl** PR/Mary Evans **59cl** IS/Richard Simpkins **59b** LOC/Alfred T. Palmer **60tl** IS/ Anthony Ladd **60tr** SS/Elena Kalistratova **60br** SS/Tom McNemar **61** PR/Andrew Lambert **62tl** IS/Anzeletti **62b** IS/Andrea Gingerich **63** IS/Sergey Kashkin

Chapter 5: Energy
64 SS/Lee Prince **65t** IS/Dane Wirtzfield **65b** SS/Cre8tive Images **66tl** IS/Leo Kin **66bl** SS/Madeline Openshaw **67bl** NASA-MSFC **68tl** IS/Lurii Konoval **68bl** SS/Chee-Onn Leong **69tr** PR/Phantatomix **69cr** PR/Tek Image **70tl** IS/Shakif Hussain **70tr** IS/Kameleon007 **70br** IS/Victor Kapas **71** IS/Jaimie D. Travis **72tl** IO/Photolibrary.com Pty., Ltd. **72tr** IO/Photos.com Select **72cr** SPL/Christian Darkin **73tl** IO/Vstock, LLC **73br** IO/FogStock, LLC **74tl** SS/ Mosista Pambudi **74br** JI **75tr** SI **75br** IO/Janos Gehring **76tl** IO/FogStock, LLC **76b** NOAA/C.Clark NOAA Photo Library, NOAA Central Library; OAR/ERL/ National Severe Storms Laboratory **77tr** Clipart.com **78tl** IS/Tamara Carter **78bl** JI **78r** JI **79tl** SS/Photomedia.com **79tr** SS/Robert Kyllo **80tl** IS/Paul McKeown **80br** IS/Ian Hamilton **81bl** PR/Martin Bond **81tr** IS/Trigga

Chapter 6: Inside the Atom
82 PR/Mike Agliolo **83t** SI/USAF **83b** IS/Adam Korzekawa **84tl** IO/Photolibrary.com Pty., Ltd. **85bl** SPL/Elscint **85tr** USGS **86tl** IS/Bjorn Kindler **87bl** IS/Emrah Turudu **87tr** SPL **88tl** JI **88r** WI **89** AP **90tl** SS/Paul Maguire **91** SPL/Tek Images **92tl** SS/Katrina Brown **92r** SS/Chris Brown **93tr** IS/Domin23 **93br** PR/Phanie **94tl** SS/R. Gino Santa Maria **94br** interactions.org **95bl** interactions.org **95br** SS/Michael Brown **96tl** IO/FogStock, LLC **97tl** IO/Photolibrary.com Pty., Ltd. **97cl** interactions.org **97tr** Interactions.org **97cr** SS/Kristian **98tl** JI **98r** LOC **99** NASA

Ready Reference
100cl PR/A. Barrington Brown **100tr** PR/Science Source **100cr** WI **101tl** PR/Mary Evans **101bl** PR/SPL **101cr** WI **101cr** NFW Service**102l** WI **102tr** WI **102br** WI **103c** PR/Science Source **103b** Clipart **104c** SS/Gnuskin Peter **104b** SS/Frank Podgorsek **105cl** SS/Piotr Przeszlo **105tr** SS/Mark Lorch **105cb** WI **106** SS/George Unger IV **107** PR/Lawrence Berkeley Lab **111t** PR/James King-Holmes **111tl** IS/Len Tillim **111bl** PR/AIP **111tr** SS/Paul Morley **111cr** JI **111br** IS/Ra-Photos

Chapter 7: Materials
114SS/Jerome Whittingham **115t** SS/PhotoCreate **115b** SS/Joshua Haviv **116tl** SS/Andraz Cerar **116tr** SS/FlorinC **116cr** SS/Holger Mette **117** SS/Pam Burley **118tl** SS/Karen Lau **118r** NGIC/Ira Block **119bl** SS/Jonathan Larson **119tr** NASA **120tl** SS/Brad Whitsitt **120bl** IS/Rick Rhay **121tl** NGIC/ Carsten Peter **121tr** IS/Tomaz Lecstek **122tl** SS/Nguyen Thai **122br** PR/ Alfred Pasieka **123tl** SS/Andrew Kerr **123br** interactions. org/Peter Grinter/NIKHEF **124tl** SS/Bonnie Watton **124bl** PR/Daivd Luzzi, University of Pennsylvania **126tl** IS/Jose Carlos Pires Pereira 127tl SS/Lenny Abbot **127bl** PR/James Holmes/Zedcor

Chapter 8: Chemistry of Life
128 PR/National Cancer Institute **129t** IS/Lidian Neeleman **129b** SS/PhotoCreate **130tl** IO/Photolibrary. com PTY, Ltd. **130bl** IO/Keith Levit Photography **131tl** IO **131bl** SS/Magdalena Szachowska **131tr** IS/Caitlin Cahill **132tl** SS/Ali Mazraie Shadi **132r** PR/Russell Kightley **133** IS/Alex Slobodkin **134tl** LANL.gov **134br** PR/Paseika **135br** LANL.gov **136tl** JI **136r** PR/Waters & A. Salic **137tr** IS/Kevin Russ **138tl** SS/Tootles **138bl** PR/Volker Steger & Christian Bardele **139t** Clipart **140tl** SS/Mircea Bezergheanu **140br** PR/SPL **141** SS/Alexei Novikov **142tl** SS/Trout55 **142bl** SS/Roger Dale Pleis **143tr** SS/Terry Reimink **143br** SS/Peggy Easterly **144tl** IS/Benoit Beauregard **144r** NGIC/Maria Stenzel **145bl** IS/Claylib **145tr** NGIC/Natalie B. Forbes

Chapter 9: Chemistry Surrounds Us
146 SS/Billy Lobo H **147t** IS/Vasko Miokovic **147b** SS/Billy Lobo H **148tl** JI **148b** IS/Eugene Bochkarev **149tl** SS/Maja Schon **149tr** SS/Max Blain **150tl** SS/Max Blain **150tr** SS/Shironina Lidiya Alexandrovna **150br** SS/Elena Elisseeva **151bl** SS/Eric Peters **151tr** SS/Kerrie Jones **152tl** SS/Scott Rothenstein **152br** WI **153tl** SS/Wayne Matthew Syvinski **153cl** SS/Andrey Zyk **154tl** SS/Telnov Oleksii **154tr** NGIC/Maria Stenzel **155** NGIC/David Barnes **156tl** SS/Howard Sandler **157tl** IO/Able Stock **157br** SS/Keir Davis **158tl** SS/Kai Hecker **158bl** SS/Martin Kucera **158r** PR/Patrick McDonnell **159bl** PR/Sunsumu Nishinaga **159tr** PR/J. Bavosi

Chapter 10: Chemistry of the Environment
160 SS/Jennifer Stone **161t** IO/Photolibrary.com Pty., Ltd. **161b** JI **162tl** NGIC/K. Yamashita/Panoramic Images **162tr** IO/FogStock **163tl** SS/Aaron Kohr **163br** IS/Alex Hinds **164tl** Amygdala Imagery **165tl** IO/DesignPics, Inc. **165br** IO VStock, LLC **166tl** SS/Norma Cornes **167bl** IS/Christopher Messer **167tr** NASA **168tl** WI **168r** JI **169** AP/Bob Child **170tl** IO/VStock, LLC **170br** SS/Steve Shoup **170r** SS/Richard Foote **171** AP/Laura Rauch **172tl** SS/Owen Hugh Retief **172tr** SS/Kharidehal Abhirama Ashwin **172br** SS/David William Taylor **173** SS/Tonis Valing

Chapter 11: Looking for Chemicals
174 PR/Geoff Tompkinson **175t** SS/Joan Ramon Mendo Escoda **175b** Clipart **176tl** WI **177** PR/Geoff Tompkinson **178tl** SS/Digitalvision **178bl** PR/AJ Photos **178br** SS/ Trout55 **179** SS/Katrina Brown **180tl** SS/Catalin Stefan **180bl** PR/Charles D. Winters **180r** WI **181** PR/Mauro Fermariello **182tl** WI **182bl** SS/WH Chow **183tr** PR/Susumu Nishinagaå **183br** PR/Hewlett-Packard Laboratories **184tl** SS/Leah-Anne Thompson **184tr** AP/Amy Sinsisterra **185** IS/Tomasz Resiak **186tl** JI **186tr** WI **186b** PR/James King-Holmes

Chapter 12: Where Do We Go From Here?
188 IO/FogStock, LLC **189t** SS/Jakub Semeniuk **189b** SS/Scott Rothstein **190tl** WI **191** WI **192tl** SS/Eyup Alp ERMIS **192bl** SS/Christian Darkin **193** PR/Franz Himpsel, University of Wisconsin **194tl** SS/Remi Cauzid **194bl** PR/Phantatomix **195** SS/Cristian Alexandru Ciobanu **196tl** WI **196tr** SPL/Steve Taylor **197** SS/Mark Lorch **198tl** SS/Anita Patterson Peppers **198r** WI **199** SPL/Sinclair Stammers **200tl** SS/Ronald Sumners **200bt** SS/BarbaraJH **201tl** PR/Kenneth Eward **201br** SS/Scott Rothenstein

At the Smithsonian
208tr SI **208bl** SI **209ct** SI **209bl** SI **209br** SI

Cover Art:
IO/Wallace Garrison PR/Alfred Pasieka